D1241371

JOHN PAUL II'S CONTRIBUTION TO CATHOLIC BIOETHICS

Philosophy and Medicine

Volume 84

CATHOLIC STUDIES IN BIOETHICS 3

JOHN PAUL II'S CONTRIBUTION TO CATHOLIC BIOETHICS

Edited by

CHRISTOPHER TOLLEFSEN

University of South Carolina, Columbia, SC, U.S.A.

A C.I.P. Catalogue record for this book is available from the Library of Congress.

ISBN 1-4020-3129-7 (HB)
ISBN 1-4020-3130-0 (e-book)

Published by Springer,
P.O. Box 17, 3300 AA Dordrecht, The Netherlands.

Sold and distributed in North, Central and South America
by Springer,
101 Philip Drive, Norwell, MA 02061, U.S.A.

In all other countries, sold and distributed
by Springer,
P.O. Box 322, 3300 AH Dordrecht, The Netherlands.

Printed on acid-free paper

springeronline.com

Printed in the Netherlands.

TABLE OF CONTENTS

CHAPTER ONE

CHRISTOPHER TOLLEFSEN

INTRODUCTION: JOHN PAUL II'S CONTRIBUTION TO CATHOLIC BIOETHICS

1. POPE JOHN PAUL II AT THE BEGINNING OF THE TWENTY-FIRST CENTURY

Any list of the most influential figures of the second half of the twentieth century would arguably have to begin with the name of Pope John Paul II. From 1978, when he was inaugurated, to the present, over a quarter of a century later, the Pope has been a dominant force in the world, both within the Catholic and Christian Church, and in the larger international community.

In the former, the Pope has spearheaded a spiritual revival of Catholicism in a Church frequently at odds with modernity, confused in the wake of the Second Vatican Council, and beset by internal divisions. The full fruits of this revival will, no doubt, not be fully known for many years; and at present they are sadly overshadowed by the recent sexual abuse crisis in the American Catholic Church. But few Catholics of the twenty-first century would fail to acknowledge the role that John Paul has played in defining for them what it means to be Catholic in this new age.

The Pope's influence has not rested within the boundaries of Catholicism, or even Christianity. Perhaps no Pope has been as committed to ecumenism, and to overcoming the divisions within Christianity that are an affront to Christ's desire that "they may all be one." Nor have previous Popes shown the remarkable sensitivity to Jews and Muslims that John Paul II has, in particular in his self-identification with the cause of atonement for the sins of Christians against their brothers and sisters of other faiths, and especially against the Jews, with whom the Pope clearly feels a deep kinship. And on the level of international politics and statesmanship, the Pope has played a significant role in the collapse of communism, especially within his native Poland, and has been a persistent voice raised for the poor and persecuted, and against the use of unreasonable violence in a troubled world.

So great and so long is the list of John Paul's accomplishments that no end of books could, and no doubt one day will be written about his role in the Church, and in the world, at the end of the second millennium. The purpose of this volume, however, is rather narrower than a full review of his various accomplishments. Among the many areas of theology, morality, and politics in which the Pope has

C. Tollefsen (Ed.), John Paul II's Contribution to Catholic Bioethics, pp. 1–6.

made a considerable contribution, one of the most widely recognized is his often controversial and always provocative work in the field of bioethics, broadly construed.

Indeed, even within the very name "bioethics" we see the root of one of the Pope's greatest concerns, a concern for human life – bios – and for the necessary ethical respect for the dignity of every human person. This concern for the person is a constant theme throughout the Pope's career, dating back to his days as priest and bishop Karol Wojtyla, and it is a concern at the root of much the Pope's work on the international stage. But it is also at the root of his continuing interest in, and contribution to, debates and challenges within the more or less academic field of bioethics: the philosophical, legal, and theological discipline that addresses issues such as abortion, euthanasia, cloning, sexual ethics, and so on. This volume is intended as a tribute to that contribution, and an attempt to articulate and understand the nature of that contribution.

2. THEMES OF JOHN PAUL'S TEACHING IN BIOETHICS

Rooted as it is in the historical teachings and traditions of the Catholic Church, John Paul's work in the area of bioethics could not fail to articulate positions and values constantly held through the Church's history. So it should come as no surprise to find that the Pope is a strong opponent of abortion, of euthanasia and assisted suicide, and of asexual reproductive techniques, nor that he, in line with more recent social teaching, supports just health-care provisions for the poor. But the Pope has made his own a number of core themes, which recur in his discussion of a variety of contested issues, in bioethics and beyond. It is this, in part, which has made the Pope's such a distinctive voice in the field of bioethics.

Foremost among the Pope's recurring concerns is, as mentioned above, his emphasis on the essential and inviolable dignity of the human person. As Luke Gormally points out, in his contribution to this volume, John Paul's understanding of human dignity is articulated in a threefold way: human dignity is connatural, existential, and definitive. Connatural dignity is that which we possess in virtue of our nature and destiny; existential dignity is that which we attain in acting uprightly; and definitive dignity is that which will come with heavenly glory. Leaving the third aside for the moment, we could perhaps summarize one major strand of the Pope's teaching by saying that he has continually called on persons to achieve their existential dignity by respecting the connatural dignity of all other human beings.

This theme of the human dignity of the person is ubiquitous in the Pope's work; it is, in consequence, a theme returned to often by almost every single contributor to this volume. Luke Gormally provides, in fact, a magisterial treatment of the notion of the Pope's thought, but its treatment is taken up again by William May, Gavin Colvert, Laura Garcia, Patrick Lee, Andrew Lustig and John Crosby in relation to a host of issues, such as abortion, contraception, euthanasia, and the physician-patient relationship. However, not only is the theme of human dignity itself of such crucial importance; it is also at the root of a variety of other recurring themes, and is linked to other crucial concerns of the Pope.

Of especial importance in the Pope's writings on human dignity is his understanding of what the human being is, and his concern for the importance of a philosophical and theological anthropology. In sharp contrast to a variety of philosophical and theological approaches of both recent and more ancient heritage, the Pope has firmly asserted as both a truth of the Church, and as a truth open to natural reason, that the human being is not a purely spiritual entity, but is a bodily reality, a true unity of spirit and body. As Patrick Lee puts it in his contribution, "we *are* bodily entities, that we are living bodies – rational animals and persons, but essentially bodies at the same time" (Lee, 2004, p. 111).

This emphasis on the bodily reality of persons, and the correlative rejection of dualisms of all sorts has a number of important consequences. For example, if human persons are human animals, then our beginning should be the beginning of the human organism that we are. In his encyclical *Evangelium Vitae*, John Paul, drawing on the Congregation for the Doctrine of the Faith's document *Donum Vitae*, draws the obvious conclusion:

> Even if the presence of a spiritual soul cannot be ascertained by empirical data, the results themselves of scientific research on the human embryo provide "a valuable indication for discerning by the use of reason a personal presence at the first appearance of a human life: how could a human individual not be a human person" (John Paul II, 1995, no. 60).

Since, then, it is always wrong intentionally to take an innocent person's life, John Paul concludes that abortion, the direct killing of the unborn, and all other forms of embryocide, are always and everywhere morally wrong.

Similarly, the Pope concludes in *Evangelium Vitae*, and elsewhere, that euthanasia, and assisted suicide are morally impermissible, conclusions which again follow from the bodily nature and the inviolable dignity, of persons. Gormally, May, Colvert, Garcia and Lustig all draw attention in their contributions to this important teaching of the Pope. Garcia and Lee additionally emphasize the importance our bodily nature has for a right understanding of sexual morality, an area of ethics deeply linked to reproductive morality. The Pope's teachings in this area are collectively known as his "Theology of the Body;" readers are encouraged to consult the collection of sermons on these topics of the same name (John Paul II, 1997).

At the same time that John Paul stresses our bodily nature, he also stresses, as he has since his early writings, the interiority and subjectivity that our lives have. Overemphasis on the objective nature of our being can lead to a depersonalization; overemphasis on subjectivity can lead to subjectivism, something very different. In his contribution reconstructing the Pope's probable views on informed consent and the physician-patient relationship, John Crosby emphasizes this phenomenologically influenced concern of John Paul with our interior, as well as our exterior reality.

Taken together, these three themes, the dignity of the person, the bodily nature of human persons, and the interiority and subjectivity of human persons, comprise the essential features of John Paul' personalism. They lead quite naturally to three further thematic emphases of the Pope's work in bioethics, namely, the close connection between freedom and morality; the theological context necessary for a proper understanding of the person; and the nature and role of suffering.

John Paul's concern for the subjectivity of the person might, as mentioned, be mistaken for a subjectivist understanding of the person. Nothing could, of course, be further from the truth; the Pope has continually objected to the modern world view which asserts that personal autonomy, unguided and unconstrained by the moral law, is at the heart of human dignity. Nonetheless, there is an element of truth to the modern emphasis on autonomy that the Pope does not ignore, and to which May, in his essay, draws attention, namely that one crucial aspect of our interiority as persons is our capacity for free choice and self-determination. But the Pope has continually taught that it is only by allowing our choices to be guided by moral truth that we are most genuinely free; our capacity to achieve existential dignity is not realized in a moral vacuum but by understanding our place in the moral order that God has created for us, for our well-being. To reject that order and choose "autonomously" in the modern way is to reject not only true freedom, but true human flourishing, a point well made in May's, Colvert's, and Shuman's essays.

What leads to the modern denial of objective moral truth, and false assertions of the sovereignty of the unguided will? An enormously important aspect of the Pope's thought is that a major source of this loss of the truth about man is the loss of the truth about God. Although the Pope strongly defends the natural law, and our access to it, as May shows, John Paul is nonetheless convinced that it is ultimately by situating our understanding of the nature of the human person and human dignity in a theological context that we can truly undergo the necessary spiritual conversion. Luke Gormally brings out the theological dimension in the Pope's teaching on human dignity, a dimension that Orthodox contributor Mark Cherry suggests needs to be made even more salient.

One area in which this theological dimension is clearly of considerable importance is in the proper understanding of suffering. Human suffering stands, in the eyes of many, as a rebuke to the Christian claims of a loving God and the message of redemption. John Paul has been at pains to provide a Christian understanding of suffering that is neither glib and sentimental, nor blind to the role that suffering plays in Christ's salvific work. Gormally, Lustig and Shuman all provide helpful accounts of the Pope's understanding of this difficult problem.

3. JOHN PAUL II AND THE GIFT

Six related themes, then, constitute the foundation of the Pope's contribution to Catholic bioethics: the dignity of the person, the bodily nature of persons, the interiority, subjectivity, and free self-constitution of persons, the connection between freedom and morality, the necessity of theology and faith, and the importance of suffering. But all six themes are in turn connected to a seventh, and for the purposes of this introduction, final core concern of Pope John Paul, the concept of the gift.

From one perspective, the concept of the gift appears to be the most fundamental concept in Judeo-Christian history and theology, and Pope John Paul has emphasized its role as perhaps no previous Christian thinker has. All creation is from its very beginning a sheerly gratuitous gift of an all-loving God, who has no need to become Creator, but does so entirely for the sake of His creation. Nor did the Creator stop there, for after the Fall of Man, God so loved the world that He gave His only son, who in turn, gave His life, that we might have life, and have it abundantly (John, 3:16, 10:10).

It from an awareness of the magnitude of this gift that John Paul has worked in articulating the themes, and defending the positions discussed above. The dignity of the human person is itself best understood in light of the gift of human life, and the gift of eternal life that is promised to us. Likewise, we are called ourselves to be gift-givers, givers of ourselves in marital communion, in the transmission of life, and in the care of others. The teachings of the Catholic Church on sex and reproduction have received new emphases from the Pope has he situates these very human activities in light of a larger, gift-centric theological context. Indeed, the entire moral life must, for John Paul, be understood in light of the gift. As Shuman quotes the Pope, "The moral life presents itself as the response due to the many gratuitous initiatives taken by God out of love for man" (John Paul II, 1993, no. 10, quoted in Shuman, 2004, p. 171). This gift-emphasis should reduce the worries of those who fear that the Pope's linkage of morality and faith, and of the moral life with Christ's revelation, is an imposing and command-based ethic.

Suffering too must be understood in light of the concept of the gift, for Christ's Passion was itself a gratuitous act of love, and in our own sufferings we are offered the opportunity to participate in Christ's redemptive act. Moreover, the presence of suffering others in our community presents us with an opportunity to make a gift of compassionate love to others, a loving gift that reverences, rather than eliminates, the life, of the other. Both Gormally and Shuman bring out these and many more points about the relationship between suffering and self-giving, both divine and human.

4. POPE JOHN PAUL II AND THE CHRISTIAN CHURCH

As mentioned, John Paul II has been as concerned as any Pope to reach out to non-Catholic denominations. In keeping with this desire, this volume contains contributions by three non-Catholic Christians, Mark Cherry, Michael Murray, and Joel Shuman. Together their essays indicate a general agreement with the Pope in his desire for a renewal of faith, morals and culture, even where these thinkers disagree with the Pope's specific approach. Two related sorts of disagreement should be here noted, as indicative of the direction that future discussion of moral theology between Catholics and non-Catholics should take.

While Pope John Paul, as mentioned already, has stressed repeatedly the need for reconversion to Christ, and the strong connections between morality and faith, he has also stressed the importance of the natural law, as the participation in the Divine Law made available to human beings through natural reason. Some Protestants, and some Orthodox, such as Cherry in his contribution, find the emphasis on natural law deeply problematic. Cherry's essay, written in response to May and Colvert, criticizes the claim that the natural law is available to us in the way the Catholic Church has taught. Murray likewise gives voice to Protestant concerns with the notion of a natural law, even as he tentatively accepts that some version of natural law ethics can provide moral guidance. And even Lustig, in his contribution, questions the extent to which demands of the natural law can be expected to be known by all.

At the same time, Murray's approach to the natural law seems more like the older natural function based tradition that Catholic thinkers, including, in this volume, Patrick Lee, have sought to revise. As befits the Pope's role, he has not put forth properly philosophical arguments, either concerning the precise extent to which he believes the natural law to be knowable, or for a particular natural law theory, whether old or new. The debate indicates, therefore, at least two areas in which Catholic discussion of the natural law with non-Catholics must, philosophically and theologically, proceed if the Pope's ecumenical hopes are to be realized.

5. ACKNOWLEDGEMENTS

It is my hope that this volume will serve as a tribute to John Paul II's contribution to field of Catholic bioethics. I would be remiss, however, in failing to acknowledge the contribution of others to this tribute. In particular, I thank the authors of the various essays in this volume for their labor and their patience; Lisa Rasmussen for her essential assistance; and H. Tristram Engelhardt for multiple forms of aid and charity.

REFERENCES

John Paul II (1995). *Evangelium Vitae*. Vatican City: Libreria Editrice Vaticana.
John Paul II (1997). *Theology of the Body*. J.S. Grabowski (Ed.), Boston: Daughters of St. Paul.
Lee, P. (2004). 'The Human Body and Sexuality in the Teaching of Pope John Paul II.' In C. Tollefsen (Ed.), *John Paul II's Contribution to Catholic Bioethics*. Dordrecth: Kluwer Academic Publishers, pp. 107-120.

CHAPTER TWO

LUKE GORMALLY

POPE JOHN PAUL II'S TEACHING ON HUMAN DIGNITY AND ITS IMPLICATIONS FOR BIOETHICS

I. JOHN PAUL II'S TEACHING ON HUMAN DIGNITY

1. INTRODUCTION

The first and larger part of this paper is expository: it aims to bring together the various elements comprising Pope John Paul II's *papal* teaching[1] on human dignity. In doing so it seeks to remain close to the textual evidence, principally in the papal encyclicals. The second part of the paper discusses in a selective fashion some of the intellectual challenges the papal teaching on human dignity poses for those seeking to develop an adequate bioethic.

Pope John Paul's understanding of human dignity is unambiguously theological in character; it is, in other words, based on divine revelation. Philosophy plays an ancillary role in the exposition of this understanding: it serves to make explicit what is implicit in revelation and also serves to articulate and defend the presuppositions of revealed truth.

In giving an account of the Pope's teaching it will help to bear in mind three distinct senses of the term 'dignity' identified in the course of Christian theological development.[2] "Dignity", Aquinas wrote, "signifies something's goodness on account of itself (*propter seipsum*)".[3] Human beings may be said to possess dignity either (1) in virtue of their nature and destiny, or (2) in virtue of the manner in which they live, or (3) in virtue of their achievement of complete fulfilment in heavenly glory. The first we could call *connatural* dignity – the sort that comes with being the kind of creatures we are; the second we could call *existential* (or acquired) dignity; and the third we could call *definitive* dignity. It is important to bear in mind that the term 'connatural' as used here should not be taken to refer simply to the ontological constitution of human beings, but includes reference to the fact that we are made for a particular fulfilment or perfection. For "we know fully what something is only when we know what it is in its final perfection"[4], a perfection to which we come through the grace of God's friendship.

C. Tollefsen (Ed.), *John Paul II's Contribution to Catholic Bioethics*, pp. 7–33.

John Paul II does not use the nomenclature introduced here, tending rather to use the term 'dignity' in an undifferentiated way, but most of what he has to say on the topic of dignity relates either to connatural or to existential dignity. Accordingly, this essay, in expounding his thought on human dignity, will make use of those two concepts.

2. THE REVELATION OF HUMAN DIGNITY

A recurrent point of reference in the Pope's teaching about human dignity is section 22 of the *Pastoral Constitution on the Church in the Modern World (Gaudium et Spes)* of the Second Vatican Council, and in particular the opening paragraphs of that section:

> In reality it is only in the mystery of the Word made flesh that the mystery of man truly becomes clear. For Adam, the first man, was a type of him who was to come, Christ the Lord. Christ the New Adam, in the very revelation of the mystery of the Father and of his love, fully reveals man to himself and brings to light his most high calling....
>
> He who is the "image of the invisible God" (*Col* 1:15), is himself the perfect man who has restored to the children of Adam that likeness to God which had been disfigured ever since the first sin. Human nature, by the very fact that it was assumed, not absorbed, in him, has been raised in us also to a dignity beyond compare (Vatican Council II, 1975, pp. 922-923).

This passage captures three ideas which are keys to the Pope's teaching about human dignity:

1. That the significance of man's creation "in the image of God" – the concept fundamental to the Christian understanding of human dignity – is made clear only in Christ's restoration in his humanity of our "likeness to God";
2. That Christ reveals man in revealing "the mystery of the Father and his love" – a revelation which is made in revealing Christ's own relationship to his Father;
3. That our dignity is further revealed in Christ's revelation of our destiny.

2.1 Connatural Dignity

Fundamental to our connatural dignity is the truth that man is made "in God's image".[5] "In procreation ... through the communication of life from parents to child, God's own image and likeness is transmitted, thanks to the creation of the immortal soul"(John Paul II, 1995, no. 43).[6] And the footnote to this statement in *Evangelium Vitae* quotes the important doctrinal statement of Pope Pius XII that "The Catholic faith requires us to maintain that souls are directly created by God" (1950, p. 575).[7] This fundamental truth about the origin of each of us "as a special gift from the Creator" contains "not only the foundation and source of the essential dignity of the human being – man and woman – in the created world, but also the beginning of the call to both of them to share in the intimate life of God himself"(John Paul II, 1988, no. 9).

Our creation not only means that we possess a distinctive kind of life but that that life is destined to have its fulfilment in a sharing in God's own life. The fullest single statement of these truths in the doctrinal teaching of Pope John Paul is probably section 34 of *Evangelium Vitae*:

> Life is always a good....*Why is life a good?* The question is found everywhere in the Bible, and from the very first pages it receives a powerful and amazing answer. The life which God gives man is quite different from the life of all other living creatures, inasmuch as man, although formed from the dust of the earth (cf. *Gen* 2:7; 3:19; *Job* 34:15; *Ps* 103:14; 104:29) *is a manifestation of God in the world, a sign of his presence, a trace of his glory* (cf. *Gen* 1:26-27; *Ps* 8:6). This is what St Irenaeus of Lyon wanted to emphasise in his celebrated definition: "Man, living man, is the glory of God". Man has been given *a sublime dignity*, based on the intimate bond which unites him to his Creator: in man there shines forth a reflection of God himself.
>
> The Book of Genesis affirms this when, in the first account of creation, it places man at the summit of God's activity, as its crown, at the culmination of a process which leads from indistinct chaos to the most perfect of creatures. *Everything in creation is ordered to man and everything is made subject to him*: "Fill the earth and subdue it; and have dominion over...every living thing" (1.28); this is God's command to the man and the woman. A similar message is found also in the other account of creation: "The Lord God took the man and put him in the Garden of Eden to till it and keep it" (*Gen* 2.15). We see here a clear affirmation of the primacy of man over things; these are made subject to him and entrusted to his responsible care, whereas for no reason can he be made subject to other men and almost reduced to the level of a thing.
>
> In the biblical narrative, the difference between man and other creatures is shown above all by the fact that only the creation of man is presented as the result of a special decision on the part of God, a deliberation to establish *a particular and specific bond with the Creator*: "Let us make man in our image, after our likeness" (*Gen* 1:26). *The life* which God offers man *is a gift by which God shares something of himself with his creature.*
>
> Israel would ponder at length the meaning of this particular bond between man and God. The Book of Sirach too recognises that God, in creating human beings, "endowed them with strength like his own, and made them in his own image" (17: 3). The biblical author sees as part of this image not only man's dominion over the world but also *those spiritual faculties which are distinctively human*, such as reason, discernment between good and evil, and free will: "He filled them with knowledge and understanding, and showed them good and evil" (*Sir* 17:7). *The ability to attain truth and freedom are human prerogatives* inasmuch as man is created in the image of his Creator, God who is true and just (cf. *Dt* 32:4). Man alone, among all visible creatures, is capable of knowing and loving his Creator" (Second Vatican Ecumenical Council, Pastoral Constitution on the Church in the Modern World *Gaudium et Spes*, 12). The life which God bestows upon man is much more than mere existence in time. It is a drive towards fullness of life; *it is the seed of an existence which transcends the very limits of time*: "For God created man for incorruption, and made him in the image of his own eternity (*Wis* 2:23)"(John Paul II, 1995, no. 34).

Here John Paul II reads the biblical witness to the meaning of man's creation 'in the image of God' as encompassing a number of truths which ground an adequate understanding of human connatural dignity:

- human life is a distinctive kind of life involving an "intimate bond" uniting each human being to his Creator (in virtue of which we have a fundamental orientation to God as our 'end').[8]
- the life of each human being is a *gift from God.*
- human beings are ends in themselves, not subordinate to things but rather with a vocation to dominion over things, and not reducible to the level of a thing (mere means) in relation to other human beings.

- our creation in the image of God means that we are endowed with fundamental capacities in virtue of which we can come to know the truth and achieve true freedom. The transcendent fulfilment of these capacities, for which we are destined, is union with God in knowledge and love.

2.1.1 Image of God and Sharing in Divine Life

John Paul II follows St Thomas Aquinas (who in turn reflects a Patristic tradition) in distinguishing between man who is "*in* (or *to*) the image of God" and Christ, the Incarnate Word, who *is* the Image of the Father. And so:

> Man created in the image of God acquires, in God's plan, a special relationship with the Word, the Father's Eternal Image, who in the fullness of time will become flesh.[9]

Man's being made "*to* the image of God" implies a fundamental orientation of his being "towards full openness to the truth" (Saward, 1995, p. 78) – to the One who is Truth in his very Person. Since the Word who is the Image is the *Son*, our orientation to the Truth is an orientation to a *filial* relationship to the Father. The proper connatural orientation of our being is one of obedience to the One who is the source of the truth of our being. We are so constituted that this orientation is to be realised in communion with others in self-giving love, for the God 'to' whose image we are made is a Trinity of Persons:

> In his intimate life, God "is love" (1 *Jn* 4:8, 16), the essential love shared by the three divine Persons: personal love is the Holy Spirit as the Spirit of the Father and the Son... It can be said that in the Holy Spirit the intimate life of the Triune God *becomes totally gift*, an exchange of mutual love between the divine Persons and that through the Holy Spirit God exists in the mode of gift (John Paul II, 1986, no. 10).

It follows that fraternity and solidarity are proper to the kind of beings we are:

> The Spirit who builds up communion in love creates between us a new fraternity and solidarity, a true reflection of the mystery of mutual self-giving and receiving proper to the Most Holy Trinity (John Paul II, 1995, no. 76).

The *capacity* for "a sincere gift of self" John Paul II regards as belonging to the very definition of a person:

> The human being is a person, a subject who decides for himself. At the same time, man "cannot fully find himself except through a sincere gift of self" [*Gaudium et Spes* 24].... this description, indeed this definition of the person, corresponds to the fundamental biblical truth about the creation of the human being – man and woman – in the image and likeness of God. This is not a purely theoretical interpretation, nor an abstract definition, for it *gives an essential indication of what it means to be human*, while emphasising *the value of the gift of self, the gift of the person* (John Paul II, 1988, no. 18).

2.1.2 The Life of Each Human Being is a Gift from God

In order to understand human dignity we need to pause to reflect on the significance of the proposition that the life of each human being is a gift from God. As we ordinarily use the word (as in "Jack gave Jill a gift for her birthday") the act of giving a gift standardly involves a recipient in a position to receive. But the recipient of the gift of life is not prior to the gift. It is the gift of life which brings a person into existence: his or her very existence is freely bestowed by God and sustained by God. We human beings are fundamentally gift: it belongs to our very nature to be free gift of God. So we live in a relationship of radical dependence on God. If we understand this we realise that we properly enjoy life on God's terms.

2.1.3 Man "Is the Only Creature on Earth that God Willed for its Own Sake"

The title of this subsection is taken from *Gaudium et Spes* no. 24, a text that John Paul II refers to frequently. God has created us with a view to our fulfilment as human persons. But this fulfilment is not meant to be an individualistic or egocentric affair.

> After affirming that man is the only creature on earth which God willed for itself, the [Second Vatican] Council immediately goes on to say that he *cannot "fully find himself except through a sincere gift of self"*. This might appear to be a contradiction, but in fact it is not. Instead it is the magnificent paradox of human existence: an existence called *to serve the truth in love* (John Paul II, 1994, no. 11).

Each human person is an end in himself and never a mere means.[10] For each of us is called to a fulfilment in that final state of beatitude in which the integrity of each will be most fully realised in a communion of self-giving and receiving through which we share in the interpersonal communion of the Trinity.

2.1.4 The Capacities Through Which Humans Image God

"The 'image of God', consisting in rationality and freedom, expresses the greatness and dignity of the human subject who is a person" (John Paul II, 1986, no. 36). We are created with a capacity for truth and a capacity for love – with reason and will. Love assumes freedom. But the exercise of our capacity for free choice – if it is to be consistent with our destiny of sharing in the life of God – must be informed by reason's grasp of truth. Our reason is so constituted as to be capable of grasping those truths which orient us to our ultimate fulfilment:

> It is the nature of the human being to seek the truth. This search looks not only to the attainment of truths which are partial, empirical or scientific; nor is it only in individual acts of decision-making that people seek the true good. Their search looks towards an ulterior truth which would explain the meaning of life. And it is therefore a search which can reach its end only in reaching the absolute. Thanks to the inherent capacities of thought, man is able to encounter and recognise a truth of this kind (John Paul II, 1998, no. 33).

Our connatural orientation to the truth is the key to the transcendence in which our existential dignity is manifested.

As we have already noted, our proper connatural orientation to the truth is one of obedience to the source of truth – the Triune God. The realisation in our lives of a fully adequate relation to the truth is 'in the Holy Spirit through the Son', whereby we come to share in the Son's own filial relationship to the Father. Human freedom is not compromised by such obedience, precisely because it is obedience to the truth:

> Patterned on God's freedom, man's freedom is not negated by his obedience to the divine law; indeed, *only through his obedience does it abide in the truth and conform to human dignity* (John Paul II, 1993, no. 42, emphasis added).

2.1.5 Human Dignity and the Body

We identify what is distinctive about human life by reference to the *exercise* of the capacities of reason and free will. But it is a mistake to think of human dignity as attaching in some exclusive way to the exercise of these capacities. Connatural dignity belongs to us in virtue of the fact that we are living human beings and, as such, persons of a particular kind. And the human person is a body/soul unity. In the Encyclical *Veritatis Splendor*, John Paul II traces a characteristic pattern of error in contemporary moral theology to the denial that fundamental aspects of the human good are to be identified by reference to what properly fulfils certain basic human, including bodily, tendencies. The denial is motivated by the belief that human freedom should not be bound by such limits; rather the body, it is proposed, is at the disposal of human freedom. In face of this pattern of error the Pope points out that:

> It contradicts the *Church's teachings on the unity of the human person,* whose rational soul is *per se et essentialiter* the form of his body. The spiritual and immortal soul is the principle of unity of the human being, whereby it exists as a whole – *corpore et anima unus* – as a person. These definitions not only point out that the body, which has been promised the resurrection, will also share in glory. They also remind us that reason and free will are linked with all the bodily and sense faculties. *The person, including the body, is completely entrusted to himself, and it is in the unity of body and soul that the person is the subject of his own moral acts.* The person, by the light of reason and the support of virtue, discovers in the body the anticipatory signs, the expression and the promise of the gift of self, in conformity with the wise plan of the Creator. It is in the light of the dignity of the human person – dignity which must be affirmed for its own sake – that reason grasps the specific moral value of certain goods toward which the person is naturally inclined. And since the human person cannot be reduced to a freedom which is self-designing, but entails a particular spiritual and bodily structure, the primordial moral requirement of loving and respecting the person as an end and never as a mere means also implies, by its very nature, respect for certain fundamental goods, without which one would fall into relativism and arbitrariness (John Paul II, 1993, no. 48).

The human body is integral to the human person and possesses the connatural dignity proper to persons. And respect for the dignity of the person entails respect for fundamental goods which are integral elements of the proper fulfilment of the person in his or her bodily reality.

2.1.6 The Distinctive Dignity of Being-a-Man and Being-a-Woman

Connatural dignity belongs to us in virtue of our created constitution as persons. But each of us is created male or female: "God created man in the image of himself, in the image of God he created him, male and female he created them" (*Gen* 1:27).

John Paul II is much concerned to elucidate the distinctive goodness (and therefore dignity) which belongs to a man in virtue of being a man and to a woman in virtue of being a woman. In the fundamental statement from *Genesis* just quoted it seems clear that the very idea of our creation "in the image of God"is elucidated by the statement "male and female he created them". The complementarity of man and woman in the sexual relationship which is marriage is meant to reflect the Triune God's own life of self-giving love. The unitive and procreative dimensions of this relationship are a central manifestation in the created order of the truth that the fulfilment of the human person is to be found in the gift of self that is open to the other. Marriage belongs to the order of creation because that covenantal relationship, in which a man and a woman commit themselves unreservedly to each other, and to any new life which may be the fruit of their self-giving love, is fundamental to God's primordial design for the transmission of human life and the flourishing of human communities.[11] The relationship between husband and wife, both in its self-giving character and in its fruitfulness, is an image of Trinitarian life. In undertaking to treat each other as irreplaceable, husband and wife affirm their equality in dignity. In their distinctive roles as husband and wife they manifest something distinctive about the dignity of being a man and being a woman.

It is difficult to clarify John Paul II's thought about the distinctive character of the *connatural* dignity of man and woman without reference to what is distinctive about their *existential* dignity. It is what is distinctive about their existential dignity which helps to identify its distinctive basis in the order of creation. So at this point it will be necessary to abandon the scheme of explaining connatural dignity independently of an explanation of existential dignity.

John Paul II, in face of the distorted understandings of the dignity of woman to be found in a variety of versions of feminism, has devoted a significant part of his papal teaching to clarifying the distinctive dignity of woman. What he has to say about woman's existential dignity is best approached by reference to what he has to say about woman's prophetic vocation, for it is the living of that prophetic vocation which exhibits woman's distinctive dignity. Women, the Holy Father says, are called to witness to "the order of love" (John Paul II, 1988, no. 29).

The 'order of love' properly

> belongs to the intimate life of God himself, the life of the Trinity. In the intimate life of God, the Holy Spirit is the personal hypostasis of love. Through the Spirit, Uncreated Gift, love becomes a gift for created persons. *Love, which is of God*, communicates itself to creatures: 'God's love has been poured into our hearts, through the Holy Spirit who has been given to us (*Rm* 5:5)' (John Paul II, 1988, no. 29).

Existentially, human persons image God in the communion of reciprocal giving and receiving which reflects that communion of love which is the Trinity. Women witness to the 'order of love' by making visible acceptance of the gift of love – fundamentally God's love – which enables them to love in return.

This witness is perhaps most readily seen in the marital relationship. At this point we need to take seriously the Pope's idea that the 'language of the body' is a clue to God's intentions in the order of creation.[12] And in doing so, it is relevant to reflect that what is distinctive of the role of the woman in marital intercourse is that she *receives* the central physical expression of her husband's love. And just in so far as she is able wholeheartedly to say 'Yes' to her husband's self-giving she is able to give herself in love and, further, accept as gift any coming-to-be of a child in her womb which may result from intercourse. That it falls to the woman to engage in a distinctive act of *receptivity* follows from the created bodily constitution of woman. So that bodily constitution itself points to what is distinctive about the *connatural* dignity of woman: "she is the one who receives love in order to love in return" (John Paul II, 1988, no. 29). And because the return of love by her establishes reciprocity "woman is the one in whom the order of love in the created world of persons first takes root" (John Paul II, 1988, no. 29). Since genuine reciprocity requires the woman's wholehearted 'Yes' to the love offered, and since authentic self-giving in marriage depends on 'God's love poured abroad in our hearts', what makes possible the woman's wholehearted 'Yes' is fundamentally God's love. Her 'Yes' is therefore a witness both to the rootedness of the human 'order of love' in the love of God and to woman's existential dignity.

This witness, according to the Pope, is not confined to marriage:

> When we say that the woman is the one who receives love in order to love in return, this refers not only or above all to the specific spousal relationship of marriage. It means something more universal, based on the very fact of her being a woman within all the interpersonal relationships which, in the most varied ways, shape society and structure the interaction between all persons – men and women (John Paul II, 1988, no. 29).

The purest expression of woman's witness to the 'order of love' is to be found in Mary:

> This *"prophetic" character of women in their femininity* finds its highest expression in the Virgin Mother of God. She emphasizes, in the fullest and most direct way, the intimate linking of the order of love – which enters the World of human persons through a Woman – with the Holy Spirit. At the Annunciation Mary hears the words: "The Holy Spirit will come upon you (*Lk* 1:35)" (John Paul II, 1988, no. 29).

The Holy Spirit came upon Mary only following her wholehearted 'Yes', her 'Fiat'. And Mary's 'Fiat' bore fruit in the conception in her womb of Jesus, the incarnate Word of God, our Redeemer.

What does John Paul II have to say about the distinctive dignity of man? Something comparatively concise: since man is complementary to woman in "*the order of love*" (John Paul II, 1988, no. 25), we can see most perspicuously in the marital relationship the distinctive dignity of man in his generative and fatherly role,[13] the dignity of the one who *loves so that the other may love*.

> In revealing and in reliving on earth the very fatherhood of God [cf. *Eph* 3:15], a man is called upon to ensure the harmonious and united development of all the members of the family...(John Paul II, 1981, no. 25).[14]

The 'order of love' of which John Paul II speaks corresponds to God's intentions in creation: to bring about, through the man-woman relationship, a communion of love which shares in and images the life of the Trinity. Human beings were made for this and are so constituted as to be capable of responding to God's love. In their constitution and calling their connatural dignity consists, a dignity which is inalienable.

2.2 Existential Dignity

Existential dignity is a matter of how human beings live; they may succeed or fail in exhibiting in their lives that dignity which depends on living in ways consistent with their connatural dignity.

At the outset of human history, by a free choice man lost his "original link with the divine source of Wisdom and Love" (*peccatum originans*), so that the condition in which we are born is one of alienation from God (*peccatum originatum*) (John Paul II, 1979, no. 8). In considering existential dignity we need to take stock of the consequences of original sin for our original vocation to live in ways consistent with our connatural dignity.

2.2.1 Existential Dignity and Human Sinfulness

In his Encyclical *Dominum et Vivificantem* (1986) John Paul II provides a profound analysis of original sin and its consequences.

We have lost our original orientation to a *filial* relationship to God, an orientation disposing us to *receive* the truth of our being from the loving source of our being and making possible an authentic exercise of freedom in self-giving love. This orientation ceased to be secure in human life because we succumbed to the lie that God, far from being the source of all that is good in our lives and of true freedom, is the enemy of man. We have been led to reject God's paternity and have fallen for the deception that our freedom (and so our dignity) depends on asserting our independence of and opposition to God:[15]

> For in spite of all the witness of creation and of the salvific economy inherent in it, the spirit of darkness is capable of showing *God as an enemy* of his own creature, and in the first place as an enemy of man, *as a source of danger and threat to man*. In this way, *Satan* manages to sow in man's soul the seed of opposition to the one who 'from the beginning' would be considered as man's enemy – and not as Father. Man is challenged to become the adversary of God!
>
> The analysis of sin in its original dimension indicates that, through the influence of the 'father of lies', *throughout the history of humanity there will be a constant pressure on man to reject God*, even to the point of hating him: 'Love of self to the point of contempt for God', as St Augustine puts it. Man will be inclined to see in God primarily a limitation of himself, and not the source of his own freedom and the fullness of good (John Paul II, 1986, no. 38).

But the loss of a right relationship to God means that "*the truth about man* becomes *falsified: who man is* and what are the *impassable limits* of his being and freedom" (John Paul II, 1986, no. 37).

Alienation from God finds its ideological expression *in the modern age* in the proclamation of the 'death of God'. But the ideology of the death of God brings with it a reductionist view of human life, manifest in contemporary anthropologies and moral theories. Human beings are seen as purely physical entities without any transcendent dimension to their existence (John Paul II, 1995, no. 22). The physicalist anthropology is matched by epistemologies – relativist and pragmatist – which deny the possibility of knowing objective truth.[16] The fundamental capacity, which lies at the root of our dignity – our capacity for knowing the truth, through the exercise of which we can find answers to those fundamental questions of meaning which naturally exercise our minds – is denied; hence the debasing nihilism widespread in our culture.[17] According to John Paul II contemporary nihilism, and the abandonment of the search for truth among philosophers "has obscured the true dignity of reason, which is no longer equipped to know the truth and to seek the absolute" (John Paul II, 1998, no. 47).

Human lives are thought of as having value just in so far as the individuals whose lives they are enjoy the rational autonomy in virtue of which they determine the value of their lives.

> Certain currents of modern thought have gone so far as to *exalt freedom to such an extent that it becomes an absolute, which would then be the source of values*. This is the direction taken by doctrines which have lost the sense of the transcendent or which are explicitly atheist (John Paul II, 1993, no. 32).

Those who lack such autonomy are all too apt to be treated as disposable:

> ...a completely individualistic concept of freedom...ends up by becoming the freedom of the 'strong' against the weak, who have no choice but to submit (John Paul II, 1995, no. 19).

The "practical materialism" of contemporary culture "breeds individualism, utilitarianism and hedonism". Since 'value' is a creation of the autonomous individual, "the body is no longer perceived as a properly personal reality, a sign and place of relations with others, with God and with the world. It is reduced to pure materiality: it is simply a complex of organs, functions and energies to be used according to the sole criteria of pleasure and efficiency" (John Paul II, 1995, no. 23).

Original sin, and our endemic "turning away from God" (John Paul II, 1986, no. 37), are, for John Paul II, the deepest root of the widespread tendency in the modern world to locate human dignity exclusively in the exercise of autonomy, to deny inherent value to bodily existence, and thereby to create a dualism of 'personal existence' and 'merely biological existence'. The consequences for human community are profoundly damaging:

> Through sin man rebels against his Creator and ends up by *worshipping creatures*: "They exchanged the truth about God for a lie and worshipped and served the creature rather than the Creator" (*Rm* 1: 25). As a result man not only deforms the image of God in his own person, but is tempted to offences against it in others as well, replacing relationships of communion by attitudes of distrust, indifference, hostility, and even murderous hatred. When God is not acknowledged *as God*, the profound meaning of man is betrayed and communion between people is compromised (John Paul II, 1995, no. 36).

2.2.2 Redemption and the Rescue of Human Dignity

Profoundly affected as we are by original sin – reason darkened[18] and freedom distorted – the achievement of existential dignity is beyond our reach. What John Paul II calls "the human dimension of the mystery of the Redemption" is the revelation to human beings of their true worth and dignity, a revelation through the manifestation of God's self-giving love for us, a love that we must allow to transform us so that the 'image of God' is restored in us and we ourselves are made free to enter into relationships of self-giving love. The image of God is restored in us through our being conformed to Christ, the Son who is the image of the unseen God, and who makes possible in us again a right relationship to God and to each other. At the beginning of his pontificate, in his first encyclical, *Redemptor Hominis* (1979), the Pope wrote:

> Man cannot live without love. He remains a being that is incomprehensible for himself, his life is senseless, if love is not revealed to him, if he does not encounter love, if he does not experience it and make it his own, if he does not participate intimately in it. This ...is why Christ the Redeemer "fully reveals man to himself". If we may use the expression, this is the human dimension of the mystery of the Redemption. In this dimension man finds again the greatness, dignity and value that belong to his humanity. In the mystery of the Redemption man becomes newly "expressed" and, in a way, is newly created. He is newly created! "There is neither Jew nor Greek, there is neither slave nor free, there is neither male nor female, for you are all one in Christ Jesus." The man who wishes to understand himself thoroughly – and not just in accordance with immediate, partial, often superficial, and even illusory standards and measures of his being – he must with his unrest, uncertainty, and even his weakness and sinfulness, with his life and death draw near to Christ. He must, so to speak, enter into him with all his own self, he must "appropriate" and assimilate the whole of the reality of the Incarnation and Redemption in order to find himself. If this profound process takes place within him, he then bears fruit not only of adoration of God but also of deep wonder at himself. How precious man must be in the eyes of the Creator, if he "gained so great a Redeemer", and if God "gave his only Son" in order that man "should not perish but should have eternal life" (John Paul II, 1979, no. 10).[19]

In order to "find himself" – in order to find again "the greatness, dignity and value that belong to his humanity" – man must "appropriate and assimilate the whole of the reality of the Incarnation and Redemption". It is clear, then, that for John Paul II existential dignity – living well in accordance with our connatural dignity – is possible only through our transformation in Christ, which makes possible our living 'in the order of love'. In the history of salvation the normative way to transformation is through our response in faith to the proclamation of the Word of God by the Church and through her sacraments, in which Christ effects the radical transformation which is to be lived out in our lives through the help of grace. We should turn to a consideration of what it is this transformation rectifies which is directly relevant to the understanding of existential dignity.

2.2.3 Transformation in Christ, Conversion and Existential Dignity

Baptism is the sacrament of conversion – of "the rebuilding of goodness in the subject" (John Paul II, 1984, no. 12). Conversion, John Paul explained in the encyclical *Redemptoris Missio*,

> is expressed in faith which is total and radical, and which neither limits nor hinders God's gift. At the same time it gives rise to a dynamic and lifelong process which demands a continual turning away from "life according to the flesh" to "life according to the Spirit" (cf. *Rom* 8: 3-13). Conversion means accepting, by a personal decision, the saving sovereignty of Christ and becoming his disciple (John Paul II, 1990, no. 46).

And in *Veritatis Splendor* the Pope explains something of what acceptance of the sovereignty of Christ means:

> *Following Christ* is not an outward imitation, since it touches man at the very depths of his being. Being a follower of Christ means *becoming conformed to him* who became a servant even to giving himself on the Cross (cf. *Phil* 2:5-8). Christ dwells by faith in the heart of the believer (cf. *Eph* 3:17), and thus the disciple is conformed to the Lord. This is the *effect of grace*, of the active presence of the Holy Spirit in us (John Paul II, 1993, no. 21).

Being "conformed to the Lord" centrally means being progressively conformed to his own relationship to his Father, that of filial obedience to His will. That is at the heart of the realisation of human existential dignity. There are a number of features of this conformity to Christ which make clear what is involved in the realisation of existential dignity.

2.2.3.1 The 'Intransitivity of Action' and Transformation in Christ

The first that should be mentioned is a presupposition of the process of conformation and transformation, namely the fundamental truth of moral psychology that through our enacted choices we shape our characters:

> Human acts...do not produce a change merely in the state of affairs outside of man but, to the extent that they are deliberate choices, they give moral definition to the very person who performs them, determining his *profound spiritual traits* (John Paul II, 1993, no. 71).

The truth that what we are most of all responsible for is what we make of ourselves, and the importance of character to human well being, are both frequently denied in secularist (particularly utilitarian) ethical theories.

2.2.3.2.1 Transformation in Christ, Conversion and the Appropriation of the Truth about Man

The second point to note about the process of transformation through which we are conformed to Christ, is that it has its beginning in conversion. Conversion, John Paul II wrote in *Dominum et Vivificantem*,

> *requires convincing of sin*; it includes the interior judgment of the conscience, and this, being a proof of the action of the Spirit of truth in man's inmost being, becomes at the same time a new beginning of the bestowal of grace and love: "Receive the Holy Spirit". Thus in this "convincing concerning sin" we discover *a double gift*: the gift of the truth of conscience and the gift of the certainty of Redemption. The Spirit of truth is the Counselor (John Paul II, 1986, no. 31).[20]

The initial 'moment' of conversion is the profound recognition of our sinfulness–our alienation from the Truth and the slavery of our wills. The concept of conscience, central to the Pope's understanding of conversion, is central also to his understanding of existential dignity. In *Dominum et Vivificantem* he writes:

> The Second Vatican Council mentioned the Catholic teaching on conscience when it spoke about man's vocation and in particular about the dignity of the human person. It

> is precisely the *conscience* in particular which determines this dignity...This capacity to command what is good and to forbid evil, placed in man by the Creator, *is the main characteristic of the personal subject.* But at the same time, "in the depths of his conscience, man detects a law which he does not impose upon himself, but which holds him to obedience."[21] The conscience therefore is not an independent and exclusive capacity to decide what is good and what is evil. Rather there is profoundly imprinted upon it *a priniciple of obedience* vis-a-vis the *objective norm* which establishes and conditions the correspondence of its decisions with the commands and prohibitions which are at the basis of human behaviour...The conscience is "the voice of God", even when man recognises in it nothing more than the principle of the moral order...(John Paul II, 1986, no. 43).

Conscience is 'determinative' of existential dignity precisely in so far as the concrete judgments of conscience on what to do and what to avoid are grounded in the objective truth about man, and in particular objective moral truth. And in acknowledging the implications of objective moral truth for one's own life one is in process of being restored to that obediential relationship to the Father in the Spirit through the Son, the source of all truth, in which God intended us to flourish. The Father's definitive Word of Truth is Jesus Christ, "and him crucified". It is through the Spirit's action in conforming us to Christ that conscience is rectified. Rectification leads to the recognition of certain truths about man and the human condition, some of which we have already discussed.

The first is that our condition is one of radical dependence on God who calls us to a praxis of self-giving love through which we are drawn into communion in the life of the Triune God.

The second is that human beings are never 'mere means'; they are ends in themselves, revealed in Jesus Christ as destined for the fulfilment of life in union with God.

The third important truth is that human beings are body/soul unities, so that bodily life is integral to personal life. Any living human body is a human person with the connatural dignity proper to a human person.

The fourth truth is that respect for persons requires respect "for certain fundamental goods" (John Paul II, 1993, no. 48), including goods which are the proper fulfilment of bodily tendencies (as marriage, for example, constitutes the good which properly fulfils the tendency to sexual union).

The fifth truth is that respect for fundamental goods in turn requires observance of exceptionless prohibitions on the choice of certain types of act which are contrary to the good of persons and therefore contrary to human dignity:

> Reason attests that there are objects of the human act which are by their nature 'incapable of being ordered' to God, *because they radically contradict the good of the person made in his image* [emphasis added]. These are the acts which, in the Church's moral tradition, have been termed 'intrinsically evil' (*intrinsece malum*): they are such *always and per se*, in other words, on account of their very object, and quite apart from the ultimate intentions of the one acting and the circumstances (John Paul II, 1993, no. 80).

But "how can obedience to universal and unchanging moral norms respect the uniqueness and individuality of the person, and not represent a threat to his freedom and dignity" (John Paul II, 1993, no. 85)? The answer to this question, according to John Paul II, lies "in the crucified Christ" who "*reveals the authentic meaning of freedom; he lives it fully in the total gift of himself* and calls the disciples to share in his freedom" (John Paul II, 1993, no. 85). He reveals it by revealing the truth about man, which is that his true freedom is exhibited in self-giving love, a self-giving love which also manifests "total obedience to the will of God". His "Resurrection from the dead is the supreme exaltation of the fruitfulness and saving power of a freedom lived out in truth" (John Paul II, 1993, no. 87). Moral absolutes do not restrict man's freedom; they merely exclude what could not possibly count as loving behaviour.

To the absolute negative norms correspond some of the most basic human rights, which, like the norms, are grounded in the good of the person.[22] Prominent among these basic rights is the right not to be unjustly killed (John Paul II, 1995, no. 57).

It is only by being conformed to Christ in our own lives that we can fully live the demands of 'moral truth'. This being the case, John Paul II regards the separation of morality from faith as a "more serious and destructive dichotomy" than the separation of freedom from truth (John Paul II, 1993, no. 88). In *Veritatis Splendor* he identifies faithful respect for moral absolutes as exhibiting the vital importance to morality of faith, understood as "a lived knowledge of Christ, a living remembrance of his commandments, and *a truth to be lived out*" (John Paul II, 1993, no. 88):

> The relationship between faith and morality shines forth with all its brilliance in the *unconditional respect due to the insistent demands of the personal dignity of every man*, demands protected by those moral norms which prohibit without exception actions which are intrinsically evil. The universality and the immutability of the moral norm make manifest and at the same time serve to protect the personal dignity and inviolability of man, on whose face is reflected the splendour of God (cf. *Gen* 9:5-6) (John Paul II, 1993, no. 88).

"Lived knowledge of Christ" is concretely the condition of "unconditional respect" for the "personal dignity of every man", a dignity protected by the moral absolutes.

The sixth truth is that the proper mode of our being is self-giving love. Christian marriage, as a commitment to self-giving love, is a paradigm of what is required for existential dignity. It also makes perspicuous what is distinctive about the dignity of man and the dignity of woman, the former lying in a 'generative' love which enables the other to love, the latter in a 'receptive' love which loves in return.

2.2.3.3 *Suffering, Transformation in Christ and Existential Dignity*

The third feature of the process of transformation through which we are conformed to Christ, and existential dignity is realised, is that transformation passes through suffering. At least some forms of human suffering are widely taken to show that the aspiration to existential dignity is doomed to defeat.

The first point to be made about suffering, following Pope John Paul, is that "at the basis of human suffering there is a complex involvement with sin":

> ...suffering cannot be divorced from the sin of the beginnings, from what St John calls "the sin of the world", from the sinful background of the personal actions and social processes in human history (John Paul II, 1984 no. 15).

Original sin left us wounded and weak in our nature (*vulneratus in naturalibus*) and the cumulative effects of personal sins have created a situation in which, at a variety of levels (including the biological), damage, pain and suffering are caused. Suffering is an ineluctable feature of our lives. Is the ideal of existential dignity that has been outlined here realisable in the face of suffering?

It is the Passion, Death and Resurrection of Christ which make it realisable: through them human suffering is linked to the order of love (John Paul II, 1984, no. 18). Just as Christ's Cross was the path to glory (Resurrection) so human suffering lived in union with Christ can become a manifestation of human dignity. Union with Christ means being united in our suffering with the love for and obedience to the Father Christ showed in his suffering. The distinctive dignity of the believing Christian who unites his sufferings with those of Christ is that of a certain proleptic participation in the power of the Resurrection: the human person is not crushed and defeated by suffering, but can continue to live in the order of love. This truth leads John Paul II to say that "Suffering, more than anything else, makes present in the history of humanity, the powers of the Redemption" (John Paul II, 1984, no. 27). In making his power known in the "weakness and emptying of self" which suffering involve (John Paul II, 1984, no. 23), God may allow us to glimpse human existential dignity pointing to definitive dignity.

> Down through the centuries and generations it has been seen that *in suffering there is concealed* a particular *power that draws a person interiorly close to Christ*, a special grace. To this grace many saints...owe their profound conversion. A result of such a conversion is not only that the individual discovers the salvific meaning of suffering but above all that he becomes a completely new person. He discovers a new dimension, as it were, of *his entire life and vocation*. This discovery is a particular confirmation of the spiritual greatness which in man surpasses the body in a way that is completely beyond compare. When this body is gravely ill, totally incapacitated, and the person is almost incapable of living and acting, all the more do interior *maturity and spiritual greatness* become evident, constituting a touching lesson to those who are healthy and normal.
>
> This interior maturity and spiritual greatness in suffering are certainly the *result* of a particular *conversion* and cooperation with the grace of the Crucified Redeemer. It is he himself who acts at the heart of human suffering through his Spirit of truth, through the consoling Spirit. It is he who transforms, in a certain sense, the very substance of the spiritual life, indicating for the person who suffers a place close to himself. *It is he* – as the interior Master and Guide – *who reveals* to the suffering brother and sister this *wonderful interchange*, situated at the very heart of the mystery of the Redemption. Suffering is, in itself, an experience of evil. But Christ has made suffering the firmest basis of the definitive good, namely the good of eternal salvation. By his suffering on the Cross, Christ reached the very roots of evil, sin and death. He conquered the author of evil, Satan, and his permanent rebellion against the Creator. To the suffering brother or sister Christ *discloses* and gradually reveals *the horizons of the Kingdom of God*: the horizons of a world converted to the Creator, of a world free from sin, a world being built on the saving power of love. And slowly but effectively, Christ leads into this world, into this Kingdom of the Father, suffering man, in a certain sense through the very heart of his suffering (John Paul II, 1984, no. 26).

Suffering borne in union with Christ under the guidance of the Holy Spirit is, then, the royal road to existential dignity, to a full return to a filial relationship to the Father, to fidelity to the truth, and to a self-giving love in which we are fulfilled.

Self-giving love should also be our response to those who suffer. As John Paul II remarks towards the close of his Apostolic Letter *Salvifici Doloris*

> At one and the same time Christ has taught man *to do good by his suffering* and *to do good to those who suffer*. In this double aspect he has completely revealed the meaning of suffering (John Paul II, 1984, no. 30).

The paradigmatic figure of the proper human response to suffering is the Good Samaritan, the one who exhibits sensitivity to the suffering of the other and offers the effective help which it is in his power to provide.[23] For the one who is being transformed by Christ, who is achieving the existential dignity which Christ makes possible, the suffering of the other is an invitation to love. In God's Providence, through the redemptive power of the Cross, the evil of suffering is turned into the opportunity for the transformation of human life:

> In the messianic programme of Christ, which is at the same time the programme *of the Kingdom of God*, suffering is present in the world in order to release love, in order to give birth to works of love towards neighbour, in order to transform the whole of human civilization into a "civilization of love" (John Paul II, 1984, no. 30).

It would be a mistake to confine our consideration of suffering to the physical suffering associated with illness and dying. As John Paul II points out: "...the first great chapter of the Gospel of suffering is written down, as the generations pass, by those who suffer persecution for Christ's sake" (John Paul II, 1984, no. 26).[24]

In a secularist milieu hostile to the truth of the Gospel, the witness of the doctor or the nurse to the truth about human dignity will often result in an hostility focused on them: in various forms they will face persecution. We should not exclude from our minds the possibility of martyrdom:

> The believer who has seriously pondered his Christian vocation, including what Revelation has to say about the possibility of martyrdom, cannot exclude it from his own life's horizon. The two thousand years since the birth of Christ are marked by the ever-present witness of the martyrs...May the People of God, confirmed in faith by the example of these true champions of every age, language and nation, cross with full confidence the threshold of the Third Millennium. In the hearts of the faithful, may admiration for their martyrdom be matched by the desire to follow their example, with God's grace, should circumstances require it (John Paul II, 1999, no. 13).

Pope John Paul does not suggest that many are called to martyrdom, but in the kind of institutional milieu which nowadays often characterises healthcare, many Christians will be called on to stand in opposition to what others take for granted, and to find themselves in consequence facing difficulties in their careers. In face of such difficulties, John Paul reminds us, the Christian is called "to a sometimes heroic commitment":

> Although martyrdom represents the high point of the witness to moral truth, and one to which relatively few people are called, there is nonetheless a consistent witness which all Christians must daily be ready to make, even at the cost of suffering and grave sacrifice. Indeed, faced with the many difficulties which fidelity to the moral order can demand, even in the most ordinary circumstances, the Christian is called, with the grace of God invoked in prayer, to a sometimes heroic commitment. In this he or she is sustained by the virtue of fortitude, whereby – as Gregory the Great teaches – one can actually 'love the difficulties of this world for the sake of eternal rewards' (John Paul II, 1993, no. 93).

In reflecting on what John Paul II says here in relation to the field of healthcare, it is clear that the witness of which he speaks is often required not just from doctors and nurses but also from patients. One need only think of what many doctors and nurses routinely expect of pregnant women, and of the pressure they put them under, to realise that patients too can be called "to a sometimes heroic commitment" and the suffering it entails.

2.2.4 Conclusion

The picture of existential dignity which emerges from the teaching of John Paul II is not that of the autonomous individual of modernity but that of Christian holiness. This teaching itself invites us to an intellectual 'metanoia' – to a radical conversion in our understanding of existential dignity. For existential dignity is not to be achieved by the illusory project of independently determining what is to count as valuable and of shaping one's life accordingly; it is to be achieved by conforming our minds and hearts to the mind and heart of Christ and thereby recovering the filial, obediential relationship to the Father through which we live in the truth. And the truth makes us free to love those whom God gives us to love. The realisation of existential dignity essentially belongs to what John Paul II calls "the order of love".

II. THE IMPLICATIONS OF JOHN PAUL II'S TEACHING ON HUMAN DIGNITY FOR BIOETHICS

3. BIOETHICS, THEOLOGY, AND THE THEME OF CHRISTIAN CONVERSION

3.1 Introduction

Though the understanding of human dignity is hardly the only requirement for the development of a bioethic it is nonetheless fundamental. John Paul II's teaching about human dignity can be read as proposing a wide range of intellectual challenges for those working in the field of Catholic bioethics. In this section I confine myself to identifying just a few of those challenges, as they bear on the fundamental orientation of the discipline and on substantive parts of its content.

Bioethics is frequently taken to be a multidisciplinary pursuit, embracing, besides philosophy and – possibly – theology, the disciplines of law, psychology, and sociology, and those humanistic disciplines which are thought conducive to sensitizing doctors, nurses and other healthcare workers to the experience, outlook and values of patients and colleagues. Without wishing to contest the relevance of other disciplines to a comprehensive understanding of bioethical issues, this paper treats bioethics as fundamentally a theological discipline which draws on philosophical resources.

The core of bioethics is theoretical reflection on what is required for a certain kind of praxis: the praxis of living well as a doctor or nurse or other healthcare worker, or as a person who needs the expertise of doctors and nurses either because of illness or some other condition (e.g. pregnancy). Patients because of their condition, doctors and nurses because of their vocation and roles, face specific kinds of choice, and the choices they make can profoundly affect whether they live well or badly as human beings. To live well as a human being is, as we have seen, to exhibit existential dignity.

John Paul II's teaching on human dignity – and particularly on existential human dignity – is concerned with some of the central requirements for any human being if he or she is to live well. And what he has to say in this connection makes clear why an adequate bioethic has to be a theological bioethic. As we noted earlier, Pope John Paul II spoke in *Veritatis Splendor* of the separation of morality from faith as a "more serious and destructive dichotomy" than the separation of freedom from truth. The fundamental reason it is more destructive is that the dependence of freedom on truth cannot be restored without Christian conversion. John Paul's analysis of the pathology of pervasive tendencies to subjectivism and nihilism in our culture finds the root of that pathology in original sin: in the assertion of a false autonomy, the rejection of a filial relationship to the Father which would acknowledge our radical dependence and be receptive to the truth and the love that come in the Holy Spirit and through the Son from the Father. The only available solution to the alienated condition we are in is a divinely revealed solution. So systematic reflection on what is required for living well as human beings is necessarily theological. Let us turn then to outlining some – just some – of the elements of a theological bioethic suggested by John Paul's teaching on human dignity.

3.2 The necessity of Christian conversion

We seek the assistance of healthcare professionals in relation to some of the most significant experiences of our lives: the transmission of human life; illness and suffering; dying and death. None of these is adequately understood outside the framework of orthodox Christian teaching. We do not adequately understand, for example, what is at issue in the bringing of a child into existence if we do not understand that the dignity of the child requires that he should be begotten through normal marital intercourse between spouses who are committed to each other in a relationship of self-giving love in which each treats the other as irreplaceable. For it is only in having the child come to be in and through the sexual expression of such a relationship that he is treated as a gift and as a person equal in fundamental dignity to his parents. So marital chastity is the condition for the parents of a right relationship to their child, a condition which not only embodies a true sense of their own dignity but also of the dignity of the child.

The practical significance of these truths is that spouses need a particular disposition of character: they need chastity. But chastity is not possible without Christian conversion, i.e. without the progressive conformation of our lives to the life of Christ through the love of God which the Holy Spirit pours into our hearts.

Parallel points can be made about the right living of illness and suffering and dying. We need faith to understand these realities aright, and beyond faith we need patience, humility and Christian hope. And none of these is to be had without the radical conversion of which John Paul II speaks.

And the heroic witness to truth – and in particular to those exceptionless moral requirements which safeguard human dignity – which doctors, nurses and patients can be called on to give and which can entail much suffering, call for a Christian faith and courage we will not have without the graces which come from entering a way of conversion.[25]

And when we consider the Good Samaritan model of the healthcare worker – the model of self-donative love in regard to those who come into one's care – it is evident that this ideal is not going to be liveable apart from the graces of inward conversion.

It seems clear, then, that the fundamental importance of Christian conversion should result in the nature and requirements for this conversion being recognised as themes at the basis of a Catholic bioethic. A Catholic bioethic has to explain what makes possible existential dignity in our lives, whether as patients, as doctors, as nurses, or as other healthcare workers. What makes it possible is the saving work of Christ, communicated to us through word and sacrament. Our communion with the Risen Christ through the Holy Spirit is what endows us with the dispositions we need to live well as human beings, to exhibit existential dignity in our lives.

There is an objection, which perhaps comes readily to mind, to the claim that the theme of Christian conversion should be basic to a Catholic bioethic. The objection is that Catholic bioethics is merely a subdiscipline of Catholic moral theology, and that the theme of Christian conversion belongs to the more general and foundational parts of moral theology rather than to a subdiscipline. The objection may be allowed to be fine in theory but to overlook the realities of practice. What I have in mind is a situation analogous to the one which largely obtains in the teaching of secular bioethics, the core of which could be deemed to be a subdiscipline of philosophical ethics. In practice foundational issues get exiguous treatment; we are all familiar with the potted versions of deontology, consequentialism and (possibly) virtue ethics which pass for a treatment of the foundations in the introductory chapter of textbooks of bioethics. In practice many students of secular bioethics – particularly doctors and nurses, the ones for whom the discipline is likely to make real differences to their lives and the lives of others – pursue the subject on an intellectually flimsy basis. Catholic bioethics risks being taught in a similarly inadequate way. So the case for making an adequate treatment of foundational topics integral to the teaching of a Catholic bioethic is pragmatic: in practice students will otherwise be left with a grossly inadequate understanding of what is required to live well as patients, doctors, nurses. And the suggestion being offered here, on the basis of Pope John Paul's teaching, is that the theme of Christian conversion needs to be given an adequate treatment in explaining the foundations of a Catholic bioethic.

4. CONVERSION AND THE 'RECTIFICATION' OF CONSCIENCE

Conversion 'rectifies' conscience, which as John Paul II says, is determinative of existential dignity (see 2.2.3.2 above). The process of rectification necessarily involves the appropriation of certain fundamental truths recognition of which underpins morally sound judgments of conscience. A Catholic bioethic should be at the service of intellectual 'metanoia', for intellectual conversion is the most difficult to achieve. Religious and moral conversion will release us from the motivation to rationalise inauthentic ways of living (a motivation easily recognisable in much contemporary philosophy), but achieving a clear understanding of the truths we need to hold may often involve a long process of dismantling false understandings and confusions.

Among the truths that should inform our conscience are those concerning connatural dignity (outlined in the whole of section 2.1 above) and those concerning moral norms, in particular exceptionless prohibitions (referred to briefly in 2.2.3.2 above). It is an implication of John Paul II's teaching on human dignity that all of these truths, in the face of contemporary challenges, call for our serious intellectual engagement in their defence.

As illustrations of the challenges that serious engagement in developing a Catholic bioethic must face I shall look briefly at two truths, duly emphasised by John Paul II in accordance with traditional Christian teaching about the nature of man.

4.1 The Doctrine of God's Direct Creation of Each Human Soul

As we saw in 2.1, Christian understanding of connatural dignity is grounded in the truth that man is made "in the image of God". In commenting on this John Paul wrote: "In procreation...through the communication of life from parents to child, God's own image and likeness is transmitted, thanks to the creation of the immortal soul"; to which statement he added a footnote reference to the doctrinal statement of Pope Pius XII that "The Catholic faith requires us to maintain that souls are directly created by God".[26] This teaching faces two challenges today, one negative in spirit, the other positive; the endeavour to meet each of them is surely important to the development of a Catholic bioethic.

The negative challenge to the teaching that each human soul is *directly* ("*immediate*", i.e. without mediation) created by God comes, even in Catholic circles,[27] from 'emergentists', and in other Christian circles from 'non-reductive physicalists', who have adopted these positions principally in response to evolutionary theory as offering the correct framework for understanding the emergence of all forms of life. For an emergentist, the appearance of human beings involves *no discontinuity* between rational and non-rational animals, but does require that the distinctive characteristics of rational life are not explanatorily reducible to 'lower level' properties. When Pope John Paul II considered the implications of evolutionary theory in an Address to the Pontifical Academy of

Sciences on October 22 1996 he rejected emergentism as incompatible with Christian belief:

> It is by virtue of his spiritual soul that the whole person possesses...dignity even in his body. Pius XII stressed this essential point: If the human body takes its origin from pre-existent living matter, the spiritual soul is immediately created by God...
>
> Consequently, theories of evolution which, in accordance with the philosophies inspiring them, consider the spirit as emerging from the forces of living matter or as a mere epiphenomenon of this matter, are incompatible with the truth about man. Nor are they able to ground the dignity of the person.
>
> With man, then, we find ourselves in the presence of an *ontological difference,* an *ontological leap*, one could say (John Paul II, 1996, p. 352, emphasis added).

John Paul II thinks that the apparent conflict, as he sees it, between what he goes on to name "ontological discontinuity" and evolutionary theory's claims of continuity in the emergence of human lives is to be resolved by distinguishing between the differing competences of philosophy and empirical science. It remains, however, that there are plenty of philosophical advocates of a version of continuity which rejects "ontological discontinuity" and they represent one important challenge for a Catholic bioethic committed to an anthropology compatible with the recognition of connatural dignity. In a culture heavily influenced by reductionist views of evolution, views which inform the rationalisation of abortion, infanticide and the killing of the senile elderly, a robust intellectual defence of ontological discontinuity has important practical significance.

The positive challenge in respect of the teaching about God's immediate creation of *each* human soul has been interestingly broached by Professor Robert Spaemann in a recent paper (Spaemann, 2001). Spaemann finds in John Paul II's statement in *Evangelium Vitae* that "The genealogy of the person is inscribed in the very biology of generation" (John Paul II, 1995, no. 43) a challenge to theology to elucidate the 'ontology of generation'. God's direct creation of the human soul is standardly thought of as an act of 'infusion' into an already generated body. But this way of envisaging God's act is inconsistent with the truth that a human being has only one soul which is the *forma corporis*; there is no living *human* body without a *rational* soul, and it is difficult to defend the view that at any stage of human development what is developing is not a *human* body. What is needed, according to Spaemann, to alter the standard 'occasionalist' picture of God's direct creation of the soul is further reflection on the difference between generation and production. Human beings cannot produce substances – they can only produce changes in natural substances. Generation differs from production precisely in bringing a new substance into being – in the case of human generation, a new human person. But though this new human person is generated by his parents he is not made by them – only God could do that. These considerations suggest that we should think of God as in a quite specific way directly involved in the act of human generation. Some such line of thought seems to lie behind John Paul II's statement that:

> God himself is present in human fatherhood and human motherhood quite differently from the way he is present in all other instances of begetting *on earth*. Indeed God alone is the source of that *image and likeness* which is proper to the human being, as it was received at Creation. Begetting is the continuation of Creation (John Paul II, 1995, no. 43).

Spaemann has drawn attention to an important task of elucidating the doctrine of God's direct creation of the human soul. He himself does not pretend to have done more than sketch the direction one might take in tackling that task.

The doctrine of God's 'direct' creation of each human soul is of fundamental importance to the understanding of human dignity and of particular importance in its implications for bioethical issues. The doctrine seems essential to a proper understanding of the life of *each* human being as a gift of God. The recognition of one's life – of one's very existence – as gift of God, is of central importance to the sense of one's dignity as inseparable from one's radical dependence on God and on His will for one's life. And the recognition of one's child's existence as a gift of God is important to a proper sense of the dignity of the child and of the non-possessive love one should have for him.

4.2 The Unity of Body and Soul in the Human Being

One response to the advocacy of emergentism and non-reductive physicalism is the revival of dualistic conceptions of the human being. Historically, modern medical science has developed in the shadow of Cartesian dualism, with its mechanistic, and consequently reductionist, understanding of the human body.[28] This has meant that the culture of modern medicine has been fundamentally hostile to the Catholic understanding of human bodily existence. And the underlying philosophical hostility goes part of the way to explaining why prevailing attitudes in clinical practice are so much at odds with Catholic attitudes to, for example, the beginning of human life and the diagnosis of death.

Historically, consistent Catholic teaching about the absolute impermissibility of abortion at any point from conception has not depended on the belief that there is an individual human being from conception.[29] Nonetheless, contemporary argument in favour of this teaching relies heavily on that belief. And that belief in turn is defended on the basis of the claim that the zygote and early embryo display a unity and developmental potential with all that is required in the way of inherent capacity for maturation into an adult human being. This amounts to the claim that the life of the embryo from its earliest developmental stages is informed by a rational human soul. The unity of the human being, *corpore et anima unus*, is crucial to the contemporary Catholic defence of the moral status of the embryo.

An understanding of human bodily existence as essentially informed by a rational soul should also be seen as having implications for controversies about the diagnosis of death, controversies of great significance for the practice of transplantation. An integrated human organism is a human person. As long as it displays organic unity it should be understood as informed by a rational soul, whether or not it is capable of exhibiting developed manifestations of human rationality. This is an implication of the teaching about the unity of the human being, a unity constituted by its being informed by a single (rational) soul.

For some time now there has been a growing body of evidence that various protocols for the diagnosis of so-called 'brain-death' (understood as establishing the death of the person) give positive results even when the human organism declared to be 'dead' exhibits integrative activity. Evidence for an integrated human organism, however deprived the condition of that organism's life, bespeaks the presence of a rational soul. So there are strong grounds for thinking that various protocols for the diagnosis of 'brain death' do not diagnose the death of the person. (Of course there are those who think that what they call 'personhood' depends on the presence of exercisable rational abilities, and who therefore conclude that diagnosis of the irreversible loss of consciousness does diagnose the death of the human being *qua* person. But this view is clearly contrary to Catholic teaching about the unity of the human being.)

Pope John Paul II has addressed these issues on a number of occasions, most recently in an address to the 18th International Congress of the Transplantation Society (John Paul II, 2000). Unfortunately, the Address assumes (a) that what the Pope calls a 'neurological criterion' of death is a recognised alternative to cardio-respiratory signs that *death has already occurred* and (b) that there is an international consensus in the scientific community both about what is required to satisfy the 'neurological criterion' of death, and about the significance of what is required. The claim that the 'neurological criterion' is an alternative sign that death has already occurred is widely contested, and there is in fact no consensus that the 'neurological criterion' requires the complete and irreversible cessation of all brain activity, nor is there consensus that such total cessation, if it occurred, would be the loss of that living integrated unity that is a human person. So the Holy Father's conclusion that clinicians, in employing conventional tests to verify the 'neurological criterion', can achieve moral certainty about the death of patients seems not well founded. One can only assume that John Paul II was not supplied with adequate information about the increasingly widespread divergences of opinion both about what is required to diagnose 'brain death', and about the significance of doing so.

It seems to be the case that there is some way to go before a Catholic anthropology has been brought to bear in an adequately informed way on what is happening in the field of transplant medicine. There is an increasing tendency to say that the initial talk in the 1960s and 1970s about 'brain death' as diagnostic of death of the person was a fiction designed to promote social acceptance of the removal of vital organs from still living patients for transplant purposes. The social acceptance having succeeded, what has triumphed, it is claimed, is the view that irreversibly unconscious patients have no more than utility value.[30] Catholic teaching needs to offer, in reference to the realities of practice, a much more coherent analysis of the ethics of transplantation consistent with maintaining the dignity and inviolability of every human life. At the basis of such an analysis there will need to be a rigorous explanation and defence of Catholic teaching about the body/soul unity of the human person, an analysis which will then need to be related to empirical data.

It should be clear, then, from these brief reflections on determining the beginning and the end of human life that a Catholic bioethic needs the support of metaphysical reflection on the constitution of the human person. But reflection on the unity of the human being is important for other bioethical concerns as well. At 2.1.5 we noted the importance of the teaching about the body/soul unity of the human person for grounding recognition that fundamental aspects of the human good are to be identified by reference to what properly fulfils certain basic human, including *bodily*, tendencies. A dualistic understanding of human persons, which understands the body in mechanistic terms, and assigns a purely instrumental value to it, is radically subversive of Christian morality, most conspicuously in relation to sexual activity. So it should be clear that, for a number of reasons, the metaphysics of the constitution of the human person has to be an integral part of an adequate Catholic bioethic.

5. CULTURAL CRITIQUE

The full-scale intellectual 'metanoia' involved in putting on the mind of Christ calls for a critical consciousness of where the Christian stands in opposition to prevalent trends and assumptions characteristic of contemporary culture. John Paul II in his writings has developed an analysis of the conflict between the culture of death and the culture of life in our society.[31] That conflict is far more basic to understanding our situation – particularly in the world of healthcare – than the conflict between C. P. Snow's 'two cultures'. So I would suggest that cultural critique should be an integral part of what a Catholic bioethic offers.

It is important both for patients and for doctors and nurses who are seeking to live well as Christians that they should be critically aware of many of the assumptions about human life and human responsibility which inform healthcare practice and the biomedical sciences. Reductionist assumptions about human life can be osmotically acquired without reflection in the process of medical education, and utilitarian assumptions about human responsibility are pervasive in the culture. People need to be helped to identify them and to acquire a critical perspective on them. Failure to achieve a critical distance often results in confusion, intellectual schizophrenia and professional lives which are relatively uninformed by Christian faith and understanding.

John Paul II's own cultural criticism has been very much in defence of human dignity. A Catholic bioethic should rise to the challenge of following his example in this as in other respects.

6. CONCLUSION

In the second part of this paper I have been able to do little more than selectively illustrate the range of the challenge that John Paul II's teaching on human dignity poses for those working in the field of Catholic bioethics. I am very conscious of how selective I have been in my account of the implications of John Paul II's teaching. But however partial the account, what has been emphasised does seem to be of some importance: elucidation of the need for Christian conversion as a foundational theme of a bioethic which recognises the distorting influence of original sin on human praxis; recognition of the need for metaphysical reflection on anthropological issues in order to defend a Christian understanding of human life; and the need for cultural critique so that we are clear-headed about the choices which face us as doctors, nurses and patients.

The Linacre Centre
London, England

***

I am grateful to Professor John Saward and the editor, Professor Christopher Tollefsen, and to my colleagues Helen Watt and Anthony McCarthy, for their critical observations on a first draft of this paper which assisted me in revising it. I am especially indebted to Davide Lees for the substantial research assistance I received from him which enabled me to commence work on this paper. None of these is to be held responsible for errors or oversights in the paper since I did not always follow their advice.

NOTES

1 I do not attempt to study the pre-papal writings of John Paul II, partly because this paper is intended as a contribution to a volume in celebration of his pontificate, and partly because his teaching as Pontiff has resulted in a very considerable *oeuvre* which it seems reasonable to think provides sufficient evidence for an analysis of his thought.

2 For some documentation see Gormally, 2002a.

3 St Thomas Aquinas, *III Sent.,* d.35, q.1, a.4, sol. lc.

4 On this 'teleological axiom' see Spaemann, 2001, p. 437.

5 See, e.g., John Paul II 1981, no. 22; 1993, no. 92; and 1995, no. 75.

6 The image received in procreation is what Aquinas calls the *imago creationis*, in virtue of which we possess connatural dignity. Our existential dignity exists in virtue of the *imago recreationis* effected by grace whereby the person "actually or habitually knows and loves God, though imperfectly. Definitive dignity exists in virtue of the *imago similitudinis* whereby the person knows and loves God perfectly "in the likeness of glory". The first image, Aquinas says, is found in all human beings; the second only in the upright; and the third only in the blessed. See St Thomas Aquinas, *Summa theologiae* I, q.93, art.4.

7 "Animas enim a Deo immediate creari catholica fides nos retinere iubet."

8 Realised through the order of grace in the Beatific Vision.

9 Discourse of 9 April, 1986 quoted in Saward, 1995, p. 78.

10 "The person can never be considered a means to an end; above all, never a means of 'pleasure'. The person is and must be nothing other than the end of every act. Only then does the action correspond to the true dignity of the person" (John Paul II, 1994, no.12).

11 See John Paul II, 1981, no. 11.

12 In the remainder of this paragraph I am rather more concrete about precisely what 'language of the body' is relevant than John Paul II is in, for example, *Mulieris Dignitatem*, but such concreteness seems to me necessary in order to clarify the Pope's teaching about the distinctive dignity of woman.

13 See John Paul II, General Audience Address 12 March 1980, reprinted in John Paul II, 1997, pp. 80-83 at p.81.

14 More fundamental than the differentiation of 'masculine' and 'feminine' is the truth that all human beings need to *receive* love from God in order to love, and for that reason may be understood to be in the position of the Bride in relation to God. The point is made in a number of places by Pope

John Paul II. Thus in *Mulieris Dignitatem* he writes: "The Bible convinces us of the fact that one can have no adequate hermeneutic of man, or of what is 'human', without appropriate reference to what is 'feminine'" (1988, no. 22). "Christ has entered this history and remains in it as the Bridegroom who 'has given himself'. 'To give' means 'to become a sincere gift' in the most complete and radical way: 'Greater love has no man than this' (*Jn* 15:13) According to this conception, *all human beings – both men and women – are called* through the Church, *to be the 'Bride' of Christ, the Redeemer of the world.* In this way, 'being the bride', and thus the 'feminine' element, becomes a symbol of all that is 'human'..."(1988, no. 25).

15 See also Rossetti, 2002, p. 118: "We can define original sin as *originated* as : the existential ignorance of God's paternity, which results in an egocentric self-constituting life and therefore mortal slavery to lust (cf. *Eph* 2: 1-3)". See the same author's 2000, especially at pp. 242-245.

16 On pragmatism, see John Paul II, 1998, no. 89.

17 On nihilism, see John Paul II, 1998, no. 90.

18 See, for example, the whole of section 22 of *Fides et Ratio* (John Paul II, 1998), in particular the statement: "The human capacity of knowing the truth was obscured by the repudiation of him who is the source and origin of truth."

19 See, from 16 years later, John Paul II 1995, no. 25.

20 On conscience, see further John Paul II 1993, no. 32; quoted also in John Paul II 1998, no. 98.

21 Internal quotation from Second Vatican Ecumenical Council, Pastoral Constitution on the Church in the Modern World *Gaudium et Spes*, 16.

22 "...civil authorities and particular individuals never have the authority to violate the fundamental and inalienable rights of the human person. In the end, only a morality which acknowledges certain norms as valid always and for everyone, with no exception, can guarantee the ethical foundation of social existence..." (John Paul II, 1993, no. 97).

23 See John Paul II, 1984, nos. 28-30.

24 Also expounded in the previous section 25.

25 One of the many intellectual challenges which a Catholic bioethic faces, and which I do not touch upon in this paper, is connected with the demand to give an 'oppositional' witness to moral truth. It is the challenge to distinguish clearly and convincingly between the kind of choice which does involve one in complicity in wrongdoing and the kind of choice in a collaborative enterprise which leaves one's moral integrity intact. A central question here, relating to the philosophy of action, concerns what falls within the scope of intention; it is a question to which there are seriously divergent answers among Catholic bioethicists.

26 See note 7.

27 For a sympathetic exposition of the claims of emergentism from a priest-philosopher see McMullin, 2000. For a Christian's espousal of non-reductive physicalism see, for example, Murphy, 2002.

28 For a brief explanation, see Gormally 2002a.

29 For the history, see Connery, 1977.

30 See chapter 1 of Singer, 1995.

31 Especially in the Encyclical *Evangelium Vitae*. For an exploration of various aspects of those two cultures see Gormally, 2002b.

REFERENCES

Connery, J. (1977). *Abortion. The Development of the Roman Catholic Perspective*. Chicago: Loyola University Press.

Gormally, L. (2002a). 'Human dignity: the Christian view and the secularist view', in J. Vial Correa and E. Sgreccia (Eds.), *The Culture of Life: Foundations and Dimensions*. Vatican City: Libreria Editrice Vaticana. 52-66.

Gormally, L. (Ed.). (2002b). *Culture of Life – Culture of Death*. London: The Linacre Centre.

John Paul II (1979). *Redemptor Hominis*. In J.M. Miller, CSB. (Ed.), (2001). *The Encyclicals of John Paul II*, 2nd Edition. Huntington, Indiana: Our Sunday Visitor Press.

John Paul II (1981). *Familiaris Consortio*. (Apostolic Exhortation). Available at http://www.vatican.va/holy_father/john_paul_ii/apost_exhortations/documents/hf_jp-ii_exh_19811122_familiaris-consortio_en.html.

John Paul II (1984). *Salvifici Doloris*. (Apostolic Letter). Available at http://www.vatican.va/holy_father/john_paul_ii/apost_letters/documents/hf_jp-ii_apl_11021984_salvifici-doloris_en.html.

John Paul II (1986). *Dominum et Vivificantem*. In J.M. Miller, CSB. (Ed.), (2001). *The Encyclicals of John Paul II*, 2nd Edition. Huntington, Indiana: Our Sunday Visitor Press.

John Paul II (1988). *Mulieris Dignitatem*. (Apostolic Letter). Available at http://www.vatican.va/holy_father/john_paul_ii/apost_letters/documents/hf_jp-ii_apl_15081988_mulieris-dignitatem_en.htm.

John Paul II (1990). *Redemptoris Missio*. In J.M. Miller, CSB. (Ed.), (2001). *The Encyclicals of John Paul II*, 2nd Edition. Huntington, Indiana: Our Sunday Visitor Press.

John Paul II (1993). *Veritatis Splendor*. In J.M. Miller, CSB. (Ed.), (2001). *The Encyclicals of John Paul II*, 2nd Edition. Huntington, Indiana: Our Sunday Visitor Press.

John Paul II (1994). 'Letter to Families.' Available at http://www.vatican.va/holy_father/john_paul_ii/letters/documents/hf_jp-ii_let_02021994_families_en.html.

John Paul II (1995). *Evangelium Vitae*. In J.M. Miller, CSB. (Ed.), (2001). *The Encyclicals of John Paul II*, 2nd Edition. Huntington, Indiana: Our Sunday Visitor Press.

John Paul II (1996). *Message to Pontifical Academy of Sciences on Evolution*. In *Origins* vol. 26, no. 22 (November 14, 1996). 349, 351-352.

John Paul II (1997). *The Theology of the Body. Human Love in the Divine Plan*. Boston: Pauline Books and Media.

John Paul II (1998). *Fides et Ratio*. In J.M. Miller, CSB. (Ed.), (2001). *The Encyclicals of John Paul II*, 2nd Edition. Huntington, Indiana: Our Sunday Visitor Press.

John Paul II (1999). *Incarnationis Mysterium*. (Bull Inaugurating the Jubilee Year). Available at http://www.vatican.va/jubilee_2000/docs/documents/hf_jp-ii_doc_30111998_bolla-jubilee_en.html

John Paul II (2000). *Address to the 18th International Congress of the Transplantation Society*. Available at http://www,vatican.va/holy_father/john_paul_.../hf_jp-ii_spe_20000829_transplants_en.htm.

McMullin, E. (2000). 'Biology and the Theology of the Human.' in P.R. Sloan (Ed.), *Controlling Our Destinies. Historical, Philosophical, Ethical and Theological Perspectives on the Human Genome Project*. Notre Dame, Indiana: Notre Dame University Press. 367-399.

Murphy, N. (2002). 'Supervenience and the downward efficacy of the mental: a nonreductive physicalist account of human action.' in R.J. Russell, N. Murphy, T.C. Meyering and M.A. Arbib (Eds.), *Neuroscience and the Person: Scientific Perspectives on Divine Action*. Vatican City State: Vatican Observatory Publications and Berkeley, California: Center for Theology and Natural Sciences. 147-164.

Pius XII (1950).*Humani Generis*. Available at http://www.vatican.va/holy_father/pius_xii/encyclicals/documents/hf_p-xii_enc_12081950_humani-generis_en.html.

Rossetti, C.L. (2000). ' "Perché diano gloria al Padre vostro celeste". Paternita divina e missione della chiesa.' *Lateranum*. 66. 235-258.

Rossetti, C.L. (2002). 'What does it mean for a Christian to be "against the world but for the world"?', in L. Gormally (Ed.), *Culture of Life – Culture of Death*. London: The Linacre Centre. 107-136.

Saward, J. (1995). *Christ is the Answer: The Christ-Centred Teaching of Pope John Paul II*. Edinburgh: T & T Clark.

Second Vatican Council (1975). *Gaudiem et Spes*. Available at http://www.vatican.va/archive/hist_councils/ii_vatican_council/documents/vat-ii_cons_19651207_gaudium-et-spes_en.html.

Singer, P. (1995). *Rethinking Life and Death. The Collapse of Our Traditional Ethics*. Oxford: Oxford University Press.

Spaemann, R. (2001). 'On the anthropology of the Encyclical *Evangelium Vitae*' in J. Vial Correa and E. Sgreccia (Eds), *Evangelium Vitae. Five Years of Confrontation with the Society*.Vatican City: Libreria Editrice Vaticana. 437-451.

CHAPTER THREE

WILLIAM E. MAY

JOHN PAUL II'S ENCYCLICAL *VERITATIS SPLENDOR* AND BIOETHICS

1. INTRODUCTION

In an article in the inaugural issue of the *National Catholic Bioethics Quarterly*, Romanus Cessario, O.P., asserted: "This short essay on method in Catholic bioethics assumes that the development of Catholic bioethics must proceed from the principles embodied in *Veritatis Splendor*"(2001, pp. 53-54). Cessario's own essay focused on the encyclical's emphasis on the complementarity between divine law and human freedom, contrasting this with the understanding of freedom found in proportionalist writers. It did not, however, seek to show how the teaching of this encyclical can and should inwardly shape "Catholic" bioethics.

Here I will attempt to show why *Veritatis Splendor*[1] is of such great importance for bioethics, whether under Catholic auspices or not. I will first center attention on John Paul's insistence that a sound morality is rooted in a sound anthropology or understanding of the human person, and the central significance of this for making true judgments and good moral choices, particularly in questions of bioethics. I will then set forth the reasons why John Paul II, with others in the Catholic tradition, considers freely chosen human acts of crucial importance to our *identity* as human persons and why, consequently, it is imperative that human persons inwardly shape their own choices and actions in accordance with the truth. His thought on the truth needed to guide human choices and actions will then be presented. The truth in question is that expressed in the "natural law," to which John Paul devotes considerable attention. A summary of his analysis of the morality of human actions will then be given along with his reasons for affirming the existence of *intrinsically evil acts*. I will then relate his moral thought to central issues of bioethics.

2. ANTHROPOLOGY AND ETHICS

Germain Grisez, after masterfully exposing the dualistic anthropology at the heart of the so-called "new morality," declared: "Christian moral thought must remain grounded in a sound anthropology which maintains the bodiliness of the person"

C. Tollefsen (Ed.), John Paul II's Contribution to Catholic Bioethics, pp. 35–50.

(1977, p. 329).[2] Joseph Cardinal Ratzinger, commenting on *Veritatis Splendor,* similarly stressed that John Paul II, opposing the kind of neo-Manicheanism underlying the claim that the magisterial teaching of the Church is "physicalistic,"[3] insists on the *bodily* character of human personhood, i.e., on an anthropology respecting the truth that human persons are unitary beings composed of body and soul (1994, pp. 16-17). Ratzinger continues by saying that for this reason John Paul II concludes that the moral theory underlying the charge of "physicalism" simply "does not," as John Paul II himself said, "correspond to the truth about man and his freedom. It contradicts the *Church's teaching on the unity of the human person,* whose rational soul is *per se et essentialiter* the form of his body….*The person, including the body, is completely entrusted to himself, and it is in the unity of body and soul that the person is the subject of his own moral acts"* (1993, no. 48).[4]

Consequently, the pope continues, the theory underlying this claim actually revives,

> in new forms, certain ancient errors [e.g., Manicheanism] which have always been opposed by the Church inasmuch as they reduce the human person to a "spiritual" and purely formal freedom. This reduction misunderstands the moral meaning of the body and of kinds of behavior involving it (cf. 1 Cor 6:19). Saint Paul declares that "the immoral, idolaters, adulterers, sexual perverts, thieves, the greedy, drunkards, revilers, robbers" are excluded from the Kingdom of God (cf. 1 Cor 6:9). This condemnation – repeated by the Council of Trent – lists as "mortal sins" or "immoral practices" certain specific kinds of behavior the willful acceptance of which prevents believers from sharing in the inheritance promised to them. In fact, *body and soul are inseparable*: in the person, in the willing agent and in the deliberate act, *they stand or fall together* (1993, no. 49).

In an important footnote appended after the words "Council of Trent" in the passage just cited, John Paul II makes it even clearer that the body and bodily life are integral to the person, for in it he calls attention to texts from both Old and New Testaments unequivocally condemning "as mortal sins certain modes of conduct involving the body" (1993, no. 49, footnote 88).

The truth that human persons are *bodily* beings and that human bodily life is a good *of* the person, intrinsic to the person, and not merely a good *for* the person, extrinsic to the person, is at the heart of a sound bioethics, Catholic or otherwise. Unfortunately, much contemporary speech about bioethics is rooted in a dualistic understanding of human beings, one sharply distinguishing between being a living human body or living individual member of the human species and being a person. John Paul II firmly opposes this dualism. He writes:

> the person, by the light of reason and the support of virtue, discovers in the body the anticipatory signs, the expression and promise of the gift of self, in conformity with the wise plan of the Creator. It is in the light of the dignity of the person – a dignity which must be affirmed for its own sake – that reason grasps the specific moral value of certain goods towards which the person is naturally inclined. And since the person cannot be reduced to a freedom which is self-designing, but entails a particular spiritual *and bodily* (emphasis added) structure, the primordial moral requirement of loving and respecting the person as an end and never as a mere means also implies, by its very nature, respect for certain fundamental goods, without which one would fall into relativism and arbitrariness (1993, no. 48).

And among these goods which must be respected is human bodily life (cf. no. 13).

The dualism underlying much contemporary bioethics regards the "person" as a conscious subject aware of itself as a self and capable of relating to other selves, i.e., other conscious subjects, and it regards the "body" as a privileged instrument of the person. On this view not every living human body, not every living member of the human species, is a person or subject of rights, but only those members of the human species who have at least incipient cognitive abilities.[5] This dualism, firmly repudiated by John Paul II in *Veritatis Splendor,* is the basis of contemporary evaluations of human actions and attitudes regarding organic human life and sexuality, as the following passage from a contemporary philosopher-theologian eloquently illustrates.

> If the person really is not his body, then the destruction of the life of the body is not directly and in itself an attack on a value intrinsic to the human person. The lives of the unborn, the lives of those not fully in possession of themselves – the hopelessly insane and the 'vegetating' senile – and the lives of those who no longer can engage in praxis or problem solving become lives no longer meaningful, no longer valuable, no longer inviolable. If the person is really not his or her own body, then the use of the sexual organs in a manner which does not respect their proper biological teleology is not directly and in itself the perversion of a good of the human person (Grisez, 1977, p. 325).

From what has been said thus far the crucial significance of an adequate anthropology of the human person for a sound bioethics should be apparent. Sound philosophy (leaving aside, for the moment, divine revelation) rejects the dualistic understanding of the human person, prevalent in contemporary Western cultures, that sharply distinguishes the conscious subject from the biologically alive body. This dualism, so rightly rejected by John Paul II in *Veritatis Splendor*, is utterly irreconcilable with the truth that the very same organism that senses, that sees, hears, smells, tastes, imagines, etc. (all of these *bodily activities*) is the very same organism that reasons, makes judgments regarding the truth and falsity of propositions, makes free choices etc. (activities attributed to the "conscious self"). This organism is *one,* not two, and this organism *is* the human person, a unity of body and soul.[6]

3. THE CRUCIAL IMPORTANCE OF FREELY CHOSEN HUMAN ACTS

Human actions (and bioethics is concerned with specific kinds of human actions) are not simply physical events in the material world that come and go, like the falling of rain or the turning of the leaves. Human actions are not things that merely "happen" to a person. They are, rather, the outward expression of a person's choice, the disclosure or revelation of that person's moral identity, of his or her *being* as a moral being. For at the core of an action, as human and personal, is a free, self-determining choice, which as such is something spiritual and abides within the person, determining the very *being* of the person.[7] We can say that a human action – i.e., a free, intelligible action, whether good or bad – is the adoption by choice of some intelligible proposal and the execution of this choice through some exterior performance. But the core of the action is the free, self-determining choice that abides in the person, making him or her *to be* the kind of person he or she is.

The significance of freely chosen human acts as self-determining is well brought out in *Veritatis Splendor.* After saying "it is precisely through his acts that man attains perfection as man," John Paul II goes on to affirm: "Human acts are moral acts because they express and determine the goodness or evil of the individual who performs them. They do not produce a change merely in the states of affairs outside of man, but, *to the extent that they are deliberate choices,* they give moral definition to the very person who performs them, determining his *profound spiritual traits*" (1993, no. 71).[8] Indeed, he says, "freedom is not only the choice for one or another particular action; it is also, within that choice, a *decision about oneself* and a setting of one's own life for or against the Good, for or against the Truth, and ultimately for or against God" (1993, no. 65).

In short, we shape our character, our identity as moral beings, by what we freely choose to do. But we are *not* free to make what we freely choose to do to be good or bad, right or wrong. And we *know* this, because we know that at times we have freely chosen to do things which we knew, at the very time we freely chose to do them, were *morally bad* – and if we claim that we have not had this experience then we are, as St. John reminds us, "liars" (see 1 Jn 1:8).

4. FREE CHOICE AND THE NEED FOR MORAL TRUTH

Our *choices,* while determining both what we will do *and* our moral identity, do not determine whether the deeds we choose to and in and through which we freely give ourselves our identity as moral beings are morally good or bad. But our choices are not blind, for they are made only after intelligent deliberation, only after *thinking,* in practical terms, about what-we-are-to-do. And if our choices are to be morally good, they must be guided by the *truth about what is to be done*. Moreover, we are capable of discovering this truth because we are intelligent beings. Indeed, we *know,* deep in our hearts, that we are called to seek the truth about what we are to do, to cleave to it once we have discovered it, and to shape our choices, our actions, and our lives in accord with it.[9]

The truth we need to help us discriminate between alternatives of action that are morally good and morally bad is practical in nature. It has to do with what we-are-going-to-do, and not with what-is. The truth in question is rooted in God's eternal law, or his wise and loving plan for human existence, and in the "natural law," which is in essence the intelligent participations of human persons in this "wise and loving plan." John Paul II treats of this truth at length in *Veritatis Splendor,* not only in his discussion of the relationship between freedom (free choice) and the truth in the first part, "Freedom and Law" (nos. 35-53), of the second chapter of *Veritatis Splendor* but in other parts of the document as well. Here I will draw together his teaching on this subject.

5. THE TEACHING OF *VERITATIS SPLENDOR* ON NATURAL LAW

Pope John Paul II affirms, with Vatican Council II,[10] that the highest norm of human action is God's divine law: eternal, objective and universal, whereby he governs the entire universe and the ways of the human community according to a plan conceived in wisdom and in love (1993, no. 43). He emphasizes that "natural law" is our intelligent participation in God's eternal law (cf. nos. 12, 40). Moreover, with St. Thomas, whom he cites extensively, particularly on this point (cf. his citation from *Summa theologiae,* 1-2, 91, 2 in no. 42), he stresses that the natural law, inasmuch as it is the participation of *intelligent, rational* creatures in God's eternal law, is properly a *human law.*[11] Thus he says, "this law is called the natural law...not because it refers to the nature of nonrational beings but because the reason which promulgates it is proper to human nature" (1993, no. 42). The moral or natural law, John Paul II affirms, *"has its origin in God and always finds its source in him."* Nonetheless, "by virtue of natural reason, which derives from divine wisdom," the natural law must also be recognized as *"a properly human law"* (1993, no. 40).

Moreover, precisely because the natural law finds its origin in God's divine and eternal law, its normative requirements are *truths* meant to help us choose rightly. In fact, John Paul II speaks of our moral life as a *theonomy,* or *participated theonomy,* since man's free obedience to God's law effectively implies that human reason and human will participate in God's wisdom and providence. By forbidding man to "eat of the tree of the knowledge of good and evil," God makes it clear that man does not originally possess such "knowledge" as something properly his own, but only participates in it by the light of the natural reason and of Divine Revelation, which manifests to him the requirements and promptings of eternal wisdom. Law must therefore be considered an expression of divine wisdom: by submitting to the law, freedom submits to the truth of creation (1993, no. 41).

John Paul II takes up the normative requirements or truths of natural law in his presentation, in chapter one of the encyclical, of the essential link between obedience to the Ten Commandments, which the Catholic tradition has always recognized as requirements of natural law, and eternal life. In his presentation of this essential link he makes it clear that the primordial moral requirement of natural law is the twofold love of God and of neighbor and that the precepts of the second tablet of the Decalogue are based on the truth that we are to love our neighbor as ourselves. This, as will be seen more clearly later, is of paramount importance.

He begins by noting that our Lord, in responding to the question posed to him by the rich young man, "Teacher, what good must I do to have eternal life?" (Mt 19:16), stresses that its answer can be found "only by turning one's mind and heart to the 'One' who is good....*Only God can answer the question about what is good, because he is the Good itself"* (1993, no. 9; cf. nos. 11, 12). He continues by saying, "God has already given an answer to this question: he did so *by creating man and ordering him* with wisdom and love to his final end, through the law which is inscribed in his heart (cf. Rom 2:15), the 'natural law.'...He also did so particularly in the "ten words," the *commandments of Sinai"* (1993, no. 12). John Paul II next reminds us that our Lord then told the young man: "If you wish to enter into life, keep the commandments" (Mt 19:17), and that Jesus, by speaking in this way, makes clear "the close connection...*between eternal life and obedience to God's commandments* [which]...show man the path of life and lead to it" (1993, no. 12). The first three of the commandments of the Decalogue call "us to acknowledge God as the one Lord of all and to worship him alone for his infinite holiness" (1993, no. 11). But the young man, replying to Jesus' declaration that he must keep the commandments if he wishes to enter eternal life, demands to know "which ones" (Mt 18:19). John Paul II says, "he asks what he must do in life in order to show that he acknowledges God's holiness" (1993, no. 13). In answering this question, Jesus reminds the young man of the Decalogue's precepts regarding our neighbor. "From the very lips of Jesus," John Paul observes, "man is once more given the commandments of the Decalogue" (1993, no. 12). These Commandments, he then affirms, are based on the commandment that we are to love our neighbor as ourselves, a commandment expressing *"the singular dignity of the human person,* 'the only creature that God has wanted for its own sake'" (1993, no. 13, with an internal citation from *Gaudium et Spes,* no. 24).[12]

It is at this point that John Paul II develops a matter of crucial importance for understanding the truths of natural law and the relationship between the primordial moral command to love our neighbor as ourselves and the specific commandments of the second tablet of the Decalogue. His point is that we can love our neighbor and respect his dignity as a person only by cherishing the real goods meant to flourish in him and by refusing to damage, destroy, or impede these goods.[13]

Appealing to the words of Jesus, John Paul II emphasizes that the different commandments of the Decalogue are really only so many reflections on the one commandment about the good of the person, at the level of the many different goods which characterize his identity as a spiritual and bodily being in relationship with God, with his neighbor, and with the material world. The commandments of which Jesus reminds the young man are meant to safeguard *the good* of the person, the image of God, by protecting his *goods* (1993, no. 13).

He goes on to say that the negative precepts of the Decalogue – "You shall not kill; You shall not commit adultery; You shall not steal; You shall not bear false witness" – "express with particular force the ever urgent need to protect human life, the communion of persons in marriage," and so on (1993, no. 13). These negative precepts, which protect the good of human persons by protecting the goods meant to flourish in them, are among the universal and immutable moral absolutes proscribing intrinsically human acts, the teaching representing, as John Paul II himself asserts, the "central theme" of the encyclical (cf. no. 115).[14]

As noted, John Paul II affirms that the negative precepts of the Decalogue, are *moral absolutes,* and that the human acts proscribed by them are *intrinsically evil acts.* It is therefore necessary, in order for us to understand properly the teaching on natural law set forth in *Veritatis Splendor*, to consider his thought on this crucially important matter, to which he devotes the fourth section of chapter two and a good part of chapter three. I will do so by examining the following points: (1) the *moral specification of human acts,* (2) the *criteria for assessing their moral goodness or badness,* (3) the truth that moral absolutes, by excluding intrinsically evil acts, *protect the inviolable dignity of human persons and point the way to fulfillment in Christ.*

6. THE MORAL SPECIFICATION OF HUMAN ACTS

John Paul II explicitly addresses this important issue in the fourth section of chapter two. After repudiating some contemporary ethical theories, which he identifies as species of "teleologism," because they are philosophically inadequate and incompatible with Catholic faith (cf. nos. 71-75), he stresses that *"the morality of the human act depends primarily and fundamentally on the 'object' rationally chosen by the deliberate will"* (1993, no. 78, with explicit reference to St. Thomas,

Summa theologiae, 1-2, 18, 6). Then, in a very important passage he writes as follows:

> In order to be able to grasp the object of an act which specifies that act morally, it is therefore necessary to place oneself *in the perspective of the acting person.* The object of the act of willing is in fact a freely chosen kind of behavior. To the extent that it is in conformity with the order of reason, it is the cause of the goodness of the will: it perfects us morally....By the object of a given moral act, then, one cannot mean a process or an event of the merely physical order, to be assessed on the basis of its ability to bring about a given state of affairs in the outside world. Rather, that object is the proximate end of a deliberate decision which determines the act of willing on the part of the acting person (1993, no. 78).

The "object" of a human act, in other words, is the subject matter with which it is concerned–it is the *intentional content* of the intelligible proposal that one adopts by choice and executes externally. For example, the "object" of an act of adultery is having intercourse with some one who is not one's spouse or with the spouse of another. This is *what* adultery is. Note that here nonmorally evaluative terms are used to describe the act in question. One is simply accurately describing precisely what the acting person is choosing to do. One is not as yet rendering a moral judgment on the act. Some people may think that the choice to do this kind of an act can be, under certain conditions, morally permissible, whereas others may think that the choice to do this kind of act is always morally bad. The "object" is simply what the person is choosing to do here and now. It is the object of his "present" intention as distinct from more remote "intentions" the agent might have in mind in making this choice: e.g., to beget a child, to obtain information needed to protect national security, etc.

7. THE CRITERIA FOR ASSESSING THE MORALITY OF HUMAN ACTS

With this understanding of the "object" of a human act in mind, it is not difficult to grasp the pope's argument, summarized as follows: "Reason attests that there are objects of the human act which are by their nature 'incapable of being ordered' to God *because they radically contradict the good of the person made in his image"* (1993, no. 80, emphasis added).

I added emphasis to this passage because it shows us that certain kinds of human acts, specified by the "object freely chosen and willed," are contrary to those precepts of natural law which prohibit acts which damage, destroy, or impede the goods perfective of human persons and in that way protect the "good" of the human person. As we saw above, John Paul II had emphasized that we can love our neighbor–the primordial moral requirement of natural law–only by cherishing and respecting the *good* of our neighbor, which we do by cherishing and respecting the *goods* perfective of him. This is the reason, as we have seen, why the precepts of the Decalogue are true requirements of natural law.

In other words, intrinsically evil acts are acts specified by the objects of intelligible proposals to damage, destroy, or impede the goods perfective of human persons. Such acts are absolutely excluded by negative precepts of natural law, moral absolutes admitting no exceptions. These precepts, moreover, do not say that it is wrong to act contrary to a virtue--e.g., to "kill *unjustly*," or "engage in *unchaste* intercourse." Rather, these precepts exclude, without exception, as John Paul II insists (cf. nos. 52, 67, 76, 82), "specific," "concrete," "particular" *kinds of behavior* (cf. nos. 49, 52, 70, 77, 79, 82) as specified by the object of human choice. Those kinds of behavior–e.g., doing something intentionally to bring about the death of an innocent person or engaging in sexual intercourse despite the fact that at least one of the parties is married–are excluded by the relevant negative moral precept without first being identified by their opposition to virtue.

As John Paul II explains, "negative moral precepts...prohibiting certain concrete actions or kinds of behavior as intrinsically evil" (1993, no. 67) protect the dignity of the person and are required by love of neighbor as oneself (1993, nos. 13; 50-52, 67, 99). Intrinsically evil acts violate (cf. no. 75) and "radically contradict" (1993, no. 80) "the good of the person, at the level of the many different goods which characterize his identity as a spiritual and bodily being in relationship with God, with his neighbor, and with the material world" (1993, no. 13; cf. nos. 78-80). It is impossible, the pope says, to respect the good of persons without respecting the goods intrinsic to them, "the goods...indicated by the natural law as goods to be pursued" (1993, no. 67), the "'personal goods'...safeguarded by the commandments, which, according to St. Thomas, contain the whole natural law" (1993, no. 79, with a reference to *Summa theologiae,* 1-2, 100, 1; cf. nos. 43, 72, 78). John Paul II emphasizes that "the primordial moral requirement of loving and respecting the person as an end and never as a mere means also implies, by its very nature, respect for fundamental goods," among which is bodily life (no. 48; cf. no. 50).

In short, according to Pope John Paul II the precepts of the Decalogue are *moral absolutes* proscribing *intrinsically evil acts.* The *truth* of these moral absolutes is rooted in the primordial principle of natural law requiring us to love our neighbors--beings who, like ourselves, are *persons* made in the image of God and who, consequently, have an inviolable dignity. These moral absolutes, required by the love commandment, protect this dignity precisely by protecting the real *goods* perfective of human persons.[15]

8. MORAL ABSOLUTES PROTECT THE INVIOLABLE DIGNITY OF HUMAN PERSONS AND POINT THE WAY TOWARD FULFILLMENT IN CHRIST

The great truth that absolute moral norms proscribing intrinsically evil acts are "valid always and for everyone, with no exception," is essentially related to the truth that human persons possess an inviolable dignity (1993, no. 97). In fact, as John Paul II observes, these norms "represent the unshakable foundation and solid guarantee of a just and peaceful human coexistence, and hence of genuine democracy, which can come into being and develop only on the basis of the equality of all its members, who possess common rights and duties. *When it is a matter of moral norms prohibiting intrinsic evil, there are no privileges or exceptions for anyone"* (1993, no. 96). To deny that there are intrinsically evil acts and moral absolutes excluding them logically leads to the surrendering of the inviolable rights of human persons, rights that must be recognized and protected if society is to be civilized.[16]

The pope recognizes "the cost of suffering and grave sacrifice...which fidelity to the moral order can demand" (1993, no. 93). Nevertheless, he takes pains to point out that the discernment which the Church exercises regarding the "teleologisms" repudiated earlier in the encyclical "is not limited to denouncing and refuting them" because they lead to a denial of moral absolutes and of intrinsically evil acts. Rather, in making this discernment the Church, in a positive way, "seeks, with great love, to help all the faithful to form a moral conscience which will make judgments and lead to decisions in accordance with the truth," ultimately with the truth revealed in Jesus (1993, no. 85). For it is *"in the Crucified Christ that the Church finds the answer"* to the question as to why we must obey "universal and unchanging moral norms" (1993, no. 85). These norms are absolutely binding because they protect the inviolable dignity of human persons, whom we are to love with the love of Christ, a self-sacrificial love ready to suffer evil rather than do it.

John Paul II illustrates this truth by appealing to the witness of martyrs. "The unacceptability of 'teleological,' 'consequentialist,' and 'proportionalist' ethical theories, which deny the existence of negative moral norms regarding specific kinds of behavior, norms which are valid without exception, is confirmed in a particularly eloquent way by Christian martyrdom" (1993, no. 90). "Martyrdom," he writes, "accepted as an affirmation of the inviolability of the moral order, bears splendid witness both to the holiness of God's law and to the inviolability of the personal dignity of man, created in God's image and likeness" (1993, no. 92), and it likewise "rejects as false and illusory whatever 'human meaning' one might claim to attribute, even in 'exceptional' conditions, to an act morally evil in itself. Indeed, it even more clearly unmasks the true face of such an act: *it is a violation of man's 'humanity'* in the one perpetrating it even before the one enduring it" (1993, no. 92, with explicit reference to *Gaudium et spes,* no. 27).

Absolute moral norms proscribing always and everywhere acts intrinsically evil by reason of the object of moral choice point the way to fulfillment in Christ, the Crucified One, who "fully discloses man to himself and unfolds his noble calling by revealing the mystery of the Father and the Father's love" (no. 92, with a citation from *Gaudium et spes,* no. 22). *"The Crucified Christ"*–who gives to us the final answer why we must, if we are to be fully the beings God wants us to be, forbear doing the evil prohibited by absolute moral norms–*"reveals the authentic meaning of freedom: he lives it fully in the total gift of himself* and calls his disciples to share in his freedom" (1993, no. 85). In a singularly important passage John Paul then writes:

> Human freedom belongs to us as creatures; it is a freedom which is given as a gift, one to be received like a seed and to be cultivated responsibly. It is an essential part of that creaturely image which is the basis of the dignity of the person. Within that freedom there is an echo of the primordial vocation whereby the Creator calls man to the true Good, and even more, through Christ's Revelation, to become his friend and to share his own divine life. It is at once inalienable self-possession and openness to all that exists, in passing beyond self to knowledge and love of the other (cf. *Gaudium et spes,* no. 24). Freedom is then rooted in the truth about man, and it is ultimately directed towards communion (1993, no. 86).

As Jesus reveals to us, "freedom is acquired in *love,* that is, in the *gift of self*...the gift of self *in service to God and one's brethren"* (1993, no. 87). This is the ultimate truth meant to guide free choices: to love, even as we have been and are loved by God in Christ, whose "crucified flesh fully reveals the unbreakable bond between freedom and truth, just as his Resurrection from the dead is the supreme exaltation of the fruitfulness and saving power of a freedom lived out in truth" (1993, no. 87).

Moreover, in our struggle to live worthily as beings made in God's image and called to communion with him–in our endeavor to shape our choices and actions in accord with the truths of natural law–we are not alone. We can live as God wills us to because he is ever ready to help us with his grace: the natural law is fulfilled, perfected, completed by the law of grace. As the Holy Father reminds us, God never commands the impossible: "Temptations can be overcome, sins can be avoided, because together with the commandments the Lord gives us the possibility of keeping them" (1993, no. 102). This truth, John Paul II points out, is a matter of Catholic faith. The Council of Trent solemnly condemned the claim "that the commandments of God are impossible of observance by the one who is justified. 'For God does not command the impossible, but in commanding he admonishes you to do what you can and to pray for what you cannot, and he gives his aid to enable you'" (1993, no. 102, citing from the Council of Trent, Session VI, Decree on Justification *Cum hoc tempore,* ch. 2; *DS* 1536; cf. Canon 18, *DS* 1568; the internal citation from Trent "For God does not command..." comes originally from St. Augustine, *De natura et gratia,* PL 44 271).

9. THE MORAL THOUGHT OF *VERITATIS SPLENDOR* AND BIOETHICAL ISSUES

John Paul II has himself related the teaching of *Veritatis Splendor* to bioethical issues. He did so in his encyclical *Evangelium Vitae,* promulgated in 1995, two years after publication of *Veritatis Splendor.* Since other essays in this volume address the relevance of *Evangelium Vitae* to bioethics, I will not consider it to any extent here. Rather, in this final part of my essay I will take up two major themes developed in *Veritatis Splendor* and show how they bear on crucially significant bioethical issues. These themes are: (1) the unity of the human person as a living organism composed of body and soul and (2) the inviolable dignity of the human person, a dignity protected by absolute moral norms or exceptionless moral norms, among them those excluding the intentional killing of innocent human beings.

With respect to the first point (1), the holistic understanding of the human person as a living organism made up of body and soul (the anthropology central to *Veritatis Splendor,* as emphasized above), I believe that it is pertinent to call attention to a significant passage in *Evangelium Vitae* where the same truth is emphasized. In the very first chapter of that document John Paul II identified as one of the root causes of the "culture of death" "the mentality which carries the concept of subjectivity to an extreme and even distorts it, *and recognizes as a subject of rights only the person who enjoys full or at least incipient autonomy and who emerges from a state of total dependence on others"* (1995, no. 19; emphasis added).

I cite this passage from *Evangelium Vitae* because it is so closely linked to the holistic anthropology at the heart of the moral thought developed in *Veritatis Splendor.* In this passage John Paul II shows that he regards as untenable and unjust the view that only certain members of the human species are the subjects of rights in the strict sense, i.e., *persons* in any morally significant sense. This view is obviously dualistic because it grants that living human bodies are indeed *biologically* identifiable as human beings or as members of the human species but holds that only those with incipient exercisable cognitive abilities – exercisable abilities of understanding, choice, and communication – must be regarded as *persons* in any meaningful sense. This view is dualistic insofar as it regards the *person* or subject with at least incipient exercisable cognitive abilities as *one* thing and the *living body* of this subject as another. As we have already seen (cf. note 5), many influential thinkers (e.g., Fletcher, Singer, Tooley) hold this view, and it is implicit in a wide variety of arguments used to justify abortion, euthanasia, killing human embryos to obtain their stem cells for research and therapeutic purposes, and other procedures central in bioethical debates.[17]

However, according to the anthropology of *Veritatis Splendor* (and of *Evangelium Vitae* as well), the human person is a unitary being composed of a spiritual element, the soul, and of a material element, the body, and *both* are integral to the *being* of the human person. What makes the body *to be human and alive* is the animating principle, the soul. But since this is so, one can say that *a living human body is a person* and that as long as we have in our midst a living human body we have in our midst a living human person, i.e., an entity intrinsically valuable, an entity that ought never to be treated merely as a means but always as an end, a being endowed with rights that are to be respected and protected by others.

Moreover (2), as we have seen in reviewing the moral thought of *Veritatis Splendor,* the inviolable dignity of the human person is protected by moral absolutes,[18] among them the absolute norm requiring us to forbear intentionally killing innocent human persons (see *Veritatis Splendor,* nos. 13, 80). And in *Evangelium Vitae* he declared: "by the authority which Christ conferred upon Peter and his Successors, and in communion with the bishops of the Catholic Church, *I confirm that the direct and voluntary [=intentional] killing of an innocent human being is always gravely immoral*" (1995, no. 57).

John Paul II also regards directly intended abortion, that is, abortion willed either as a means or an end, as an instance of the direct and voluntary killing of an innocent human being – of a person (cf. 1993, no. 80; 1995, no. 58). Moreover, he clearly thinks that an individual human being is in existence from the time of conception/fertilization (cf. 1995, no. 60). He does not, however – nor is it his responsibility as pope – to provide *arguments* to support this claim, although he does refer to scientific research on the human embryo as valuable for indicating that a new human individual is present from conception on (ibid.). In short, while rejecting, both in *Veritatis Splendor* and in *Evangelium Vitae,* the claim that only those members of the human species who have incipient exercisable cognitive abilities (or who manifest conscious abilities of some kind) can be considered *persons*, he does not seek to provide a philosophical critique of this claim or to offer philosophical reasons to support his claim that every living human being is a person.

Many excellent scholars and philosophers, however, have done this. Hence here I will summarize major arguments offered by such scholars because the *personhood,* and not the *humanity* of the biological organism in question is the issue heatedly debated in arguments regarding the killing of this organism in abortion and "involuntary euthanasia," the destruction of human embryos in order to obtain their stem cells for research and/or therapeutic procedures on *other* human subjects, and similar issues.

The reasoning behind the claim that only those members of the human species, i.e., those living bodies identifiable as *human*, who possess at least incipient exercisable cognitive abilities are persons, is fallacious. It fails to distinguish between a *radical* capacity or ability and a *developed* capacity or ability. A radical capacity, one rooted in the *being* of the entity in question, can also be called (as it is by authors like Patrick Lee) an *active,* as distinct from a merely *passive*, potentiality. An unborn or newborn human baby, precisely by reason of its nature as a human being and therefore a member of the human species, has the *radical capacity or active potentiality* to discriminate between true and false propositions, to make choices, and to communicate rationally. But in order for this human being to *exercise* this capacity or set of capacities, his radical capacity or active potentiality for engaging in these activities--which are after all predictable kinds of behavior for *human* beings or member of the human species – must be allowed to develop. But it could never be developed if it were not present to begin with. Similarly, human beings older than zygotes, embryos, fetuses (e.g., prepubescent children, teenagers, adults, senior citizens) may, because of accidents or illness, no longer be capable of exercising their capacity or ability to engage in these activities, but this in no way means that they no longer have the radical capacity or active potentiality for doing so. They are simply inhibited by disease or accidents from exercising this capacity.

In short, *a living human body* (alive and human, be it recalled, because its animating principle is the human or spiritual soul), no matter what its size (a zygote, a preimplantation embryo, a fetus, a newborn, some senile individual) has the radical capacity or active potentiality to do what human persons are supposed to do. A human zygote, embryo, newborn has the active potentiality or radical capacity to develop *from within its own resources* all it needs to exercise the property or set of properties characteristic of adult human beings. A human embryo, as philosophers Robert and Mary Joyce so precisely put matters, is a *person with potential,* not a *potential person* (1971, p. 123).

Those thinkers, who, like Singer and Tooley, require that an entity have exercisable cognitive abilities, recognize that the unborn have the *potentiality* to engage in such activities. But they consider this merely a *passive* potentiality (or do not make the distinction) and fail to recognize the critically significant difference between an *active* potentiality and a merely passive one. In his excellent development of the significance of this difference, Patrick Lee makes two very important points. The first concerns the *moral* significance of the distinction between an active and a passive potentiality. An active potentiality means "that the same entity which possesses it is the same entity as will later exercise that active potentiality. With a passive potentiality, that is not so; that is, the actualization of a passive potentiality often produces a completely different thing or substance" (1997, p. 26).

Lee's second key point is that the proper answer to the query "why should higher mental functions or the capacity or active potentiality for such functions be a trait conferring value on those who have it" is that such functions and the capacity for them are "of ethical significance not because [these functions] are the only intrinsically valuable entities but because entities which have such potentialities are intrinsically valuable. And, *if the entity itself is intrinsically valuable, then it must be intrinsically valuable from the moment that it exists"* (1997, pp. 26-27; emphasis added). As a group of British thinkers have also pointedly noted, in criticizing this alleged criterion of personhood for its arbitrariness, "it is true that the distinctive dignity and value of human life *are manifested* in those specific exercises of developed rational abilities in which we achieve some share in such human goods as truth, beauty, justice, friendship, and integrity. But the necessary rational abilities are acquired in virtue of an underlying or radical capacity, *given with our nature as human beings,* for developing precisely those abilities" (Gormally et al, 1994, pp. 123-124; emphasis in original).[19]

I cannot here enter into a discussion of the time when a living human body, i.e., a living human person, first comes into existence. I simply affirm here that the vast majority of human persons first come to be at fertilization/conception, with a tiny minority (e.g., one or more monozygotic twins, triplets, etc.) coming to be shortly thereafter as a result of a kind of "cloning." Arguments and evidence supporting this position are abundant and are perhaps best set forth by Lee (1997, chapters 1 and 2), Grisez (1989, pp. 27-47; 1993, pp. 488-505), Benedict Ashley and Albert Moraczewski (1994, pp. 33-60), Angelo Serra and Roberto Colombo (1998, pp. 128-177), Ramón Lucas Lucas (1998, pp. 178-205), May (2000, pp. 156-170), and I refer readers to these sources.

From this one can easily conclude that the anthropology and moral philosophy/theology rooted therein which we find in *Veritatis Splendor* holds as utterly immoral the choice intentionally to kill innocent human beings, no matter what their stage of development or the quality of their lives. It thus condemns, as other writings of John Paul II and ecclesiastical documents issued with his approval and authority clearly show, the following: abortion chosen as either means or end (=direct or intentional abortion) (cf. John Paul II, 1995, nos. 58-62), euthanasia or mercy killing (1995, nos. 64-65), using human embryos as "research material" or as providers of organs or tissues for transplants to other persons (1995, no. 63), the killing of human embryos to obtain their stem cells for research and/or therapeutic use on *other* human subjects (cf. John Paul II's "Remarks to President George W. Bush on Stem Cell Research").[20] It also rules out "making" babies in the laboratory by artificial insemination, *in vitro* fertilization, or cloning insofar as such procedures violate the dignity of unborn babies in the earliest stage of their genesis by treating them as products of technology, products in principle inferior to their producers and subject to quality controls and not as persons equal in nature and dignity to their parents (cf. Congregation for the Doctrine of the Faith, *Donum Vitae* [Instruction on Respect for Human Life in Its Origins and on the Dignity of Procreation], 1987). It holds that one can rightly refuse medical treatments for oneself or for those for whom one has care *if* objective reasons show that the *treatments* in question are either useless or excessively burdensome. But it utterly condemns forgoing treatments because one judges that the human life in question is useless or excessively burdensome. It does so because human life is a precious and incalculably valuable good and human life, no matter how burdened, is always a precious gift from God (John Paul II, 1995, no. 65).

10. CONCLUSION

This essay, I hope, has shown the very significant relevance to bioethics of the moral thought advanced by John Paul II in his encyclical *Veritatis Splendor* and of the anthropological understanding of the human person in which this thought is rooted.

John Paul II Institute for Studies on Marriage and Family at
The Catholic University of America
Washington, DC, USA

NOTES

1 See John Paul II, 1993. Literature commenting on *Veritatis splendor* is enormous. Among books and essays in English offering favorable and helpful studies of its moral teaching are the following: DiNoia and Cassario, 1999, a collection of essays by DiNoia and Cessario, Servais Pinckaers, O.P., Alasdair MacIntyre, Russell Hittinger, Avery Dulles, Livio Melina, Martin Rhonheimer, and William May; Finnis and Grisez, 1994; William E. May, "*Veritatis Splendor:* An Overview of the Encyclical," *Communio* 21 (1994) 228-251 [enlarged and reprinted as Chapter Eight of my 2003), pp. 269-294]. One of the most hostile attacks on the encyclical is provided by the essays in Selling and Jans, 1994, with essays by Selling, Jans, Gareth Moore, O.P., Bernard Hoose, Louis Janssens and others. In my essay, "*The Splendor of Accuracy:* How Accurate?" (1995) I show how these authors have distorted and misrepresented the teaching of John Paul II. One of the most helpful presentations of the teaching of this encyclical is given by Dionigi Tettamanzi (formerly a professor of moral theology and now Cardinal Archbishop of Milan, Italy) in his "Guida alla lettura" (1993), pp. 5-56.

2 See also Grisez, 2001.

3 This claim is commonly made by Catholic theologians who reject Church teaching on such issues as contraception, abortion, euthanasia. Two representative essays are: Curran, 1969; and Louis Janssens, 1969.

4 In a perceptive and thoughtful essay, Brian Johnstone, C.Ss.R., briefly but cogently shows that a central principle of the moral teaching of *Veritatis Splendor* and of the new *Catechism of the Catholic Church* is man's nature as a "person in the unity of soul and body." See his 1994.

5 It is this understanding of the "person" one finds in large measure in the writings of such authors as Joseph Fletcher, Michael Tooley, and Peter Singer. See, e.g., Fletcher, 1976; Tooley, 1983; Singer, 1994; Reiman, 1997.

6 Among excellent works to consult on this issue are the following: Adler, 1968; Braine, 1996; Lee, 1998.

7 The Scriptures, particularly the New Testament, are very clear about this. Jesus taught that it is not what enters a person that defiles him or her; rather, it is what flows from the person, from his or her heart, from his or her choice (see Matt 15:10-20; Mk 7:14-23).

8 At this point in the text John Paul II cites a remarkable passage from Saint Gregory of Nyssa's *De Vita Moysis*, II, 2-3: *PG* 44, 327-328: "All things subject to change and to becoming never remain constant, but continually pass from one state to another, for better or worse...Now, human life is always subject to change; it needs to be born ever anew....But here birth does not come about by a foreign intervention, as is the case with bodily beings...; it is the result of free choice. Thus *we are* in a certain way our own parents, creating ourselves as we will, by our decisions" (cited in *Veritatis Splendor,* no. 71).

9 "It is in accordance with their dignity that all men, because they are persons, that is, endowed with reason and free will, and therefore bearing personal responsibility, are both impelled by their nature and bound by a moral obligation to seek the truth....They are also bound to adhere to the truth once they come to know it and to direct their whole lives in accord with the demands of truth." Vatican Council II, Declaration on Religious Liberty, *Dignitatis Humanae,* no. 2. The Council merely echoes what Plato's Socrates died for centuries ago.

10 See Vatican Council II, Declaration on Religious Liberty, *Dignitatis Humanae,* no. 3:

11 For St. Thomas's teaching on natural law see Finnis, 1998, esp. chs. III and IV, pp. 56-131; see also my 1994, pp. 43-60.

12 Here John Paul's thought reminds us of the teaching of St. Thomas on the Decalogue. Thomas regarded the Decalogue as "proximate conclusions" from the "first and common" precepts of natural law, and he explicitly identified the two precepts commanding us to love God and neighbor as the "first and common precepts of natural law" to which all precepts of the Decalogue must be referred as conclusions are referred to their common principles. See *Summa theologiae,* 1-2, 100, 3, ad 1. This is hardly surprising, for both the Old Testament (Dt 6:5 and Lev. 19:18), and the New Testament (Matt 22:37-39, Mk 12:28-34; Lk 10:25-28) all teach this.

13 Here John Paul's thought again echoes that of St. Thomas, who not only taught that the twofold commandment of love of God and of neighbor is the first principle of morality (cf. *Summa theologiae,* 1-2, 100, 3, ad 1) but also that we offend God only by acting contrary to our own *good* (cf. *Summa contra gentiles,* 3, ch. 122).

14 In chapter one John Paul II also emphasizes that natural law, whose specific normative requirements have also been revealed in the "ten words" given on Sinai and reaffirmed by the lips of Jesus himself, is ultimately fulfilled and perfected only as *"a gift of God:* the offer of a share in the divine Goodness revealed and communicated in Jesus" (no. 17).

15 For a reasoned philosophical defense and articulation of the affirmation of moral absolutes see Finnis, 1991.

16 On this see Finnis, 1981, pp. 223-226.

17 See, for example, the reasoning used in the defenses of abortion provided by the following authors: Warren, 1973; Harrison, 1983, pp. 187-231, and the literature cited there; see also authors referred to in note 5.

18 John Finnis clearly describes the meaning of the moral absolutes or exceptionless moral norms central to the moral thought of *Veritatis Splendor*: These norms, he writes, "have the following characteristic: The types of action they identify are specifiable, as potential objects of choice, without reliance on any evaluative term which presupposes a moral judgment on an action. Yet this evaluative specification enables moral reflection to judge that the choice of any such act is to be excluded from one's deliberation and one's action…they [moral absolutes] are exceptionless in an interesting way. Exceptions to them are logically possible, and readily conceivable, but are *morally* excluded….Of all these norms the following is true: Once one has precisely formulated the type, one can say that the norm which identifies each chosen act of that type as wrong is true and applicable to every such choice, whatever the (further) circumstances. An exceptionless norm is one which tells us that, whenever we are making a choice, we should never choose to do *that* sort of thing (indeed should never even deliberate about whether or not to do it)" Finnis, 1991, pp. 2-4).

19 On this point see also the arguments marshaled in Grisez and Boyle, 1979, pp. 218-238. See also the arguments and evidence advanced by Grisez, 1989, and 1993, pp. 488-498.

20 The text of these remarks can be found in John Paul II, 2001.

REFERENCES

Adler, M. (1968). *The Difference of Man and the Difference It Makes.* New York: Meridian Books.

Ashley, B.A., O.P., and A. Moraczewski, O.P. (1994). 'Is the Biological Subject of Human Rights Present from Conception?' in P. Cataldo and A. Moraczewski (Eds.), *The Fetal Tissue Issue: Medical and Ethical Aspects* (pp. 33-60). Braintree, MA: The Pope John XXIII Medical Moral Center.

Braine, D. (1996). *The Human Animal.* Notre Dame, IN: University of Notre Dame Press.

Curran, C. (1969). 'Natural Law and Contemporary Moral Theology,' in C.E. Curran (Ed.), *Contraception: Authority and Dissent* (pp. 151-175). New York: Herder & Herder.

Cessario, R., O.P. (2001). 'Toward an Adequate Method for Catholic Bioethics,' *National Catholic Bioethics Quarterly* 1, 51-62.

DiNoia, J.A., O.P. and R. Cessario, O.P. (1999). *Veritatis Splendor and the Renewal of Moral Theology.* Chicago: Midwest Theological Forum.

Finnis, J. (1981). *Natural Law and Natural Rights.* New York and Oxford: Oxford University Press.

Finnis, J. (1991). *Moral Absolutes: Tradition, Revision, and Truth.* Washington, D.C.: The Catholic University of America Press.

Finnis, J. (1998). *Aquinas: Moral, Political, and Legal Theory.* New York and Oxford: Oxford University Press.

Finnis, J. and G. Grisez (1994). "Negative Moral Precepts Protect the Dignity of the Human Person," *L'Osservatore Romano*. English ed. No. 8, 6-7.
Fletcher, J. (1976). *The Ethics of Genetic Controls: Ending Reproductive Roulette*. New York: Doubleday.
Grisez, G. and J. Boyle, Jr. (1979). *Life and Death with Liberty and Justice: A Contribution to the Euthanasia Debate*. Notre Dame, IN: University of Notre Dame Press.
Grisez, G. (1977). 'Dualism and the New Morality,' in M. Zalba (Ed.), *L'Agire Morale,* (pp. 323-330) vol. 5 of *Atti del Congresso Internazionale (Roma-Napoli 17/24 aprile 1974) Tommaso d'Aquino nel suo Settimo Centenario*. Edizioni Domenicane Italiane. Napoli.
Grisez, G. (1989). 'When Do People Begin?' in *Proceedings of the American Catholic Philosophical Association* 63, 27-47.
Grisez, G. (1993). *Living a Christian Life*. Quincy, IL: Franciscan Press.
Grisez, G. (2001). 'Bioethics and Christian Anthropology,' *NationalCatholic Bioethics Quarterly* 1, 33-40.
Gormally, L. (Ed.) (1994). *Euthanasia, Clinical Practice and the Law*. London: The Linacre Centre.
Harrison, B.W. (1983). 'Evaluating the Act of Abortion: The Debate About Fetal Life,' in her *Our Right to Choose*. Boston: Beacon Press.
Janssens, L. (1969). 'Considerations on *Humanae Vitae*,' *Louvain Studies* 2, 231-253.
John Paul II (1993). *Veritatis Splendor*. Vatican City: Libreria Editrice Vaticana.
John Paul II (1995). *Evangelium Vitae*. Vatican City: Libreria Editrice Vaticana.
John Paul II (2001). 'Remarks to President George W. Bush on Stem Cell Research,' *National Catholic Bioethics Quarterly* 1, 617-618.
Johnstone, B., C.Ss.R. (1994). 'Personalist Morality for a Technological Age: *The Catechism of the Catholic Church* and *Veritatis Splendor*,' *Studia moralia* 32 (1994) 121-136.
Joyce, M. and R. (1971). *Come, Let Me Be Born*. Franciscan Herald Press. Chicago.
Lee, P. (1997). *Abortion and the Unborn Child*. Washington, D.C.: The Catholic University of America Press.
Lee, P. (1998). 'Human Beings Are Animals,' in R.P. George (Ed.), *Natural Law and Moral Inquiry: Ethics, Metaphysics, and Politics in the Work of Germain Grisez*. (pp. 135-151). Washington, D.C.: Georgetown University Press.
Lucas, R.L. (1998). 'The Anthropological Status of the Human Embryo,' in J. de D. Vial Correo and E. Sgreccia (Eds.), *The Identity and Statute (sic) of the Human Embryo: Proceedings of the Third Plenary Session of the Pontifical Academy for Life (Vatican City, February 14-16 1997)* (pp. 178-205). Vatican City: Libreria Editrice Vaticana.
May, W.E. (1995). '*The Splendor of Accuracy:* How Accurate?' *The Thomist* 59(3), 467-483.
May, W.E. (2000). *Catholic Bioethics and the Gift of Human Life*. Huntington, IN: Our Sunday Visitor.
May, W.E. (2003). *An Introduction to Moral Theology*. (2nd ed.) Huntington, IN: Our Sunday Visitor.
Ratzinger, J. Cardinal (1994). 'Perche un'enciclica sulla morale? Riflessioni circa la genesi e l'elaborazione della *Veritatis splendor*,' in G. Russo (Ed.), *Veritatis Splendor: Genesi, Elaborazione, Significato* (pp. 9-20). Roma: Edizioni Dehoniane.
Reiman, J. (1997). *Critical Moral Liberalism: Theory and Practice*. New York: Oxford University Press.
Selling, J.A., and J. Jans (Eds.) (1994). *The Splendor of Accuracy: An Examination of the Assertions Made in the Encyclical Veritatis Splendor*. Kampen/Grand Rapids: Kok Pharos/Eerdmans.
Serra, A. and R. Colombo (1998). 'Identity and Status of the Human Embryo: The Contributions of Biology,' in J. de D. Vial Correo and E. Sgreccia (Eds.), *The Identity and Statute (sic) of the Human Embryo: Proceedings of the Third Plenary Session of the Pontifical Academy for Life (Vatican City, February 14-16 1997)* (pp. 128-177). Vatican City: Libreria Editrice Vaticana.
Singer, P. (1994). *Rethinking Life and Death: The Collapse of Our Traditional Ethics*. New York: St. Martin's Press.
Tettamanzi, D. (1993). 'Guida alla lettura' in *Lettera Enciclica di S.S. Papa Giovanni Paolo II Veritatis Splendor: Testo integrale con introduzione e guida alla lettura di S.E. Mons. Dionigi Tettamanzi* (pp. 5-56). Casale Monferrato: Edizioni Piemme.
Tooley, M. (1983). *Abortion and Infanticide*. New York: Oxford University Press.
Warren, M.A. (1973). 'On the Moral and Legal Status of Abortion,' *The Monist* 57(1), 43-61.

CHAPTER FOUR

GAVIN T. COLVERT

LIBERTY AND RESPONSIBILITY: JOHN PAUL II, ETHICS AND THE LAW

1. INTRODUCTION

> Archbishop Wojtyla, as one of the proponents of a new *Declaration on Religious Freedom* (now styled, in Latin, *Dignitatis Humanae*), spoke in the first days of the debate, sharpening the point he had made in the third session [of Vatican II] on the relationship between freedom and truth. It was not sufficient, he argued, to say simply, "I am free." Rather, "it is necessary to say … 'I am responsible.' … Responsibility is the necessary culmination and fulfillment of freedom (Weigel, 2001, pp. 164-5).

John Paul II is an enigma to many of his contemporaries, who have received their intellectual formation from the Western liberal tradition. On the one hand, he is an ardent supporter of the religious, political and economic freedoms championed by liberal democracies, and a vigorous opponent of communism and socialism. Recent papal documents celebrate democracy's role in protecting human rights, limited government, and the institution of private property.[1] Prior to his election as Pope, he was instrumental in producing important Vatican II documents, including declarations concerning the right to religious freedom and the essential dignity of the human person.[2] Over the years he has also been a supporter of ecumenism and a patron of the sciences. Under his direction, the Vatican has sponsored dialogues between theologians and scientists, revisiting the question of the Church's handling of the Galileo affair, and even acknowledging evidence for the theory of evolution.[3] Without a doubt, he has sought to bring the Church into a vigorous dialogue with modern science and culture.

On the other hand, many liberals see John Paul as a reactionary figure. Nowhere is this perceived tension more evident than in regard to his treatment of biomedical questions and related human life issues. He has been unequivocal in his defense of the Church's teachings concerning human sexuality, which were clarified in *Humanae Vitae*, the controversial document of his predecessor Paul VI. Papal encyclicals such as *Veritatis Splendor* and *Evangelium Vitae* contain vigorous criticism of what he calls the "culture of death." This culture endorses abortion, euthanasia, assisted suicide, and destructive research with developing embryos, among other things. Proponents of these practices view them as consistent with liberal democracy's call

C. Tollefsen (Ed.), John Paul II's Contribution to Catholic Bioethics, pp. 51–72.

for greater human liberty, the right to privacy and freedom of conscience. The Pope's defense of the "culture of life" by contrast rests upon his commitment to the priority of the common good and the existence of objective moral absolutes that impose responsibilities upon the exercise of our freedom. These absolutes are embodied in the fundamental principles of the natural law. Contrary to the minimalist procedural approach of many contemporary liberal theorists, he endorses a substantive or thick conception of the common human good. This has a significant impact upon the legal and moral orders. Positive law can be criticized by more fundamental moral principles, upon which it depends for its justification. Natural human rights are not merely basic postulates. They are requirements of practical reason, reflecting a carefully articulated philosophical anthropology. Put succinctly, John Paul thinks that society and law have a duty to promote the common good in addition to protecting individual liberties. The common good is defined by a more comprehensive conception of human flourishing than liberal theorists are prepared to entertain.

One may wonder whether the Pope's moral conservatism can be reconciled with his enthusiasm for many aspects of liberal democracy. More specifically, can his appeal to certain moral absolutes be compatible with democratic principles concerning the role of law and the exercise of human freedom? This chapter answers that question affirmatively by examining some key documents from John Paul's pontificate and placing them in the context of his philosophical and theological development. Other chapters discuss how John Paul's thought leads to practical conclusions about biomedical issues such as abortion, euthanasia and embryonic stem cell research. The primary objective of the present study is to show how these views are consistent with his endorsement of democratic principles protecting liberty and rights. Some constructive criticisms of the Pope's thought from within his own tradition rightly caution that Christianity and contemporary liberal democracy are not perfectly mated philosophical viewpoints. The opposite charge that he has gone too far towards embracing modern liberal theory must also be examined.

2. JOHN PAUL: PERSONALISM AND 'THE LAW OF THE GIFT'

The underlying principles that provide coherence to John Paul's thought on law, democracy and morality can be found in his particular synthesis of philosophical personalism and Thomistic ethics, especially his "law of the gift," which stresses that human subjectivity is fundamentally relational. As a young scholar, Karol Wojtyla sought to integrate classical Thomistic natural law theory and Aristotelian metaphysics with the personalism of Max Scheler. He attempted to make use of the good in phenomenological analyses of the subject, while offering a corrective to their deficiencies. This corrective required appropriate attention to the objective order and classical metaphysics.[4] His argument was that ultimately, one could only find meaning in the depths of human subjectivity by self-transcendence and recognition of one's participation in an objective order of other human beings.[5] Wojtyla's deep faith life, especially his meditation upon the gratuitous loving action of Christ, and the unqualified loving acceptance by Mary of her role in the economy of salvation, contributed to this philosophical conclusion.

We can see the basic insight of his thought at work in his 1976 Sermon *Sign of*

Contradiction, which asserts his "law of the gift."[6] According to this principle, human beings must recognize that life itself is a gift and that our very dignity as persons originates with the gratuitous act of God's creation. That dignity is preserved and brought to fulfillment through the further free gift of Christ's incarnation for the sake of our redemption. We become most fully human and more fully like God when we imitate that gift in giving ourselves to others through responsible and loving actions.[7]

The influence of personalism is evident in John Paul's central insight. He stresses the importance of the person of Christ, God's gratuitous action, and our human fulfillment as the key data to be understood. Remarkably, he uses this insight about human subjectivity to break through a distorted Kantian sense of autonomy that predominates in the modern age, in order to reveal the basic correctness of the traditional call to self-transcendence and neighbor love. When we recognize that our dignity as persons is bestowed as a gift, we are better able to understand the transcendent source of the moral order and structure of the universe. Kantian 'autonomism', as John Paul labels it, involves a mistaken understanding of 'subjectivity' and freedom that sets these two in opposition to the natural law.[8] Whereas much contemporary philosophy is unable to get beyond the subjective dimension of experience, for John Paul the deepest self-knowledge leads to philosophical realism and Thomistic metaphysics, and the quest for self-perfection leads to the love of neighbor and concern for the common good, rather than to a misguided preoccupation with one's own liberty.[9]

The Pope is remarkably successful in the task of bringing together the phenomenological and personalistic treatment of human 'subjectivity' with the traditional emphasis upon the common good and objective natural law principles. Although the basic features of his viewpoint have been in place since before his election as Pope, his ability to balance the various dimensions of his thought has matured over time. This can be demonstrated by examining a select number of papal documents that show the progression in his thinking.

3. FIRST WORDS: *REDEMPTOR HOMINIS*

> Jesus Christ meets the man of every age, including our own, with the same words, "You will know the truth, and the truth will make you free." These words contain both a fundamental requirement and a warning: the requirement of an honest relationship with regard to the truth as a condition for authentic freedom, and the warning to avoid every kind of illusory freedom, every superficial unilateral freedom, every freedom that fails to enter into the whole truth about man and the world (John Paul II, 1979, no. 12).

So writes John Paul II in *Redemptor Hominis*, the first encyclical of his pontificate, promulgated in 1979, shortly after his election as Pope. In 1979 John Paul was already aware of various crises facing the Church and society and the need to bring about revitalization and renewal. His interest in engaging the modern world in dialogue with Christianity required a delicate synthesis between tradition and the embrace of modern philosophical and theological concepts.

If one theme can be said to unify the letter, it is the sustained reflection upon the gratuitous gift of Christ's incarnation and redemptive suffering. Without a doubt, the encyclical bears the stamp of the Pope's personalism with its focus upon the encounter of each human being with Christ's redemptive act, rather than upon abstract theological formulae. Some phenomenologists reject the postulation of an objective ontological and moral order. This is not an adequate account of the Pope's synthesis of Thomism and personalistic phenomenology. While he frequently uses Max Scheler's language of 'values', for instance, he is critical of Scheler's inability to make the transition from the subjective experience of 'value' to the objectivity of moral norms.[10] It is therefore a mistake to view the personalistic bent of the work as standing in opposition to concern for an objective moral order.

In no. 6 the Pope draws our attention to a decline in confidence concerning the truth of moral principles, and the ill effects this growing relativism is producing (see also no. 16). In several places, such as no. 17, the notion of duty, and especially of our duty toward the common good is mentioned. Most significantly, the Pope stresses that truth is a *sine qua non* for the genuine exercise of freedom. In later encyclicals, especially *Veritatis Splendor*, the primacy of truth takes on special significance in the Pope's call for a restoration of objective moral norms and voluntary self-limitation against present tendencies towards moral relativism and radical individualism. We can thus trace these important themes back to the origin of his pontificate.

On the whole, however, the emphasis of the document is upon rights and freedoms rather than upon duty and law. The Pope's awareness of a crisis within democratic institutions has clearly deepened over time. In no. 12 where he discusses the Church's mission, he focuses upon Vatican II's *Declaration on Religious Freedom*, *Dignitatis Humanae*. Both *Dignitatis Humanae* and *Redemptor Hominis* ground the right to religious freedom in a duty based in law, the duty to seek the truth in light of the command of divine law.[11] In the crucial no. 17, however, which expands and develops the concept of the responsible exercise of freedom, very little is said concerning the objective basis of rights in law and moral obligation. In fact, the language of 'natural law' is mentioned only once in passing in the section, and not again throughout the whole document.

What is mentioned repeatedly is the existence of "inviolable human rights." John Paul also speaks enthusiastically about the United Nations, expressing confidence in the future prospects for the U.N. to establish and protect "man's objective and inalienable rights" (John Paul II, 1979, no. 17). Most interestingly, whereas *Dignitatis Humanae* spoke of the state's responsibility to the common good as consisting in a concern for the rights *and duties* of the human person, no. 17 speaks only of the state serving the common good when all citizens are assured of their *rights*. He concludes that the protection of human rights is the measure by which a society's commitment to social justice can be evaluated.

This kind of language in papal documents has encouraged conservative critics within the tradition to argue that John Paul has gone too far toward embracing liberal political theory.[12] Without a set of guiding principles such as those of the natural law to direct them, delegates to various U.N. conferences have approved a raft load of new rights concerning abortion, cloning and other sensitive biomedical issues. As John Paul argues more forcefully in later encyclicals, we can only get beyond the impasse concerning spurious rights claims and unacceptable exercises of freedom when liberty is grounded in the conception of an objective moral order.[13] It would be incorrect to argue that the priority of law and duty to rights goes unrecognized in *Redemptor Hominis,* but it does remain in the background.

4. IMPORTANT DEVELOPMENTS: *DONUM VITAE*

The 1987 instruction from the Vatican Congregation for the Doctrine of the Faith, *Donum Vitae*, demonstrates a clear shift towards the reintegration of the language of natural law, duty and the priority of the common good into a more central place in the papal position. John Paul was not the author of the document, but he officially received it and authorized its publication. With frequent references to his addresses and writings, it is essentially a compilation of positions he and his immediate predecessors have taken concerning the issues. In addition, its solemn nature and the fact that it was issued on the feast of the Chair of St. Peter all point to the close association of the document with his viewpoint.[14]

The express purpose of the instruction is to foster greater respect for human dignity, including the right to life of each person, especially the human embryo.[15] The language of rights is evident in the document, but so too is the language of duty and obligation. Certain uses of the term '*ius*' are unintelligible if one takes them to mean only the modern understanding of subjective rights as moral powers, possessions or benefits, and not also as imposing burdens in justice.[16] The concept of personal dignity is closely tied to that of the gratuitous gift of life by the Creator.[17] This recalls John Paul's fundamental philosophical and theological insight, 'the law of the gift'. The immense value of the gift of life reminds us that we must take responsibility for it. Experimentation upon and manipulation of embryonic life threatens to reduce it to the status of a commodity or possession, and places the manipulator in a state of undue dominion over it. This unjust state is characterized in terms of violation of the rights of the human person, but it is clear that the document's key concern is that these rights impose correlative duties.

The introduction argues that discovery of the 'truth about man' requires an anthropology which indicates the influence of the traditional concern for an objective moral order. The study of human nature rests upon the natural law, which "expresses and lays down the purposes, rights and duties which are based upon the bodily and spiritual nature of the human person" (Congregation for the Doctrine of the Faith, 1987, no. 3). Whereas *Redemptor Hominis* had all but left unmentioned the concept of natural law as a basis for the working out of the notion of human dignity, *Donum Vitae* makes significant use of it.[18]

Two other themes from *Donum Vitae* merit attention: the duty of civil society to protect and defend the common good, and the discussion of the 'rights and duties' of spouses. As John Paul had argued in *Redemptor Hominis*, so *Donum Vitae* asserts that a primary duty of civil society is to foster the common good through the protection of basic human rights. The text stresses that the promotion of public morality requires that civil law conform to the requirements of the moral or natural law. An especially pertinent case in point, given the transformation of the term '*ius*' in modern rights theories, is the example of how spousal rights relate to the use of reproductive technologies. *Donum Vitae* argues that civil law has no authority to grant the permissibility of the use of reproductive techniques that distort or remove the rights inherent to spouses:

> Respect for the unity of marriage and for conjugal fidelity demands that the child be conceived in marriage; the bond existing between husband and wife accords the spouses, in an objective and inalienable manner, the exclusive right to become father and mother solely through each other (Congregation for the Doctrine of the Faith, 1987, II.2).[19]

This formulation cannot be reduced to the modern conception of 'rights'. Marriage provides certain liberties to spouses, but it also imposes the responsibility of marital chastity. This burden is not imposed arbitrarily, but for the sake of the common good of the family. The 'right' of a spouse here is inseparable from the notions of duty, teleology and a substantive conception of the moral good.[20] *Donum Vitae* adds that the 'right' of a spouse, emphatically does not include the "right to have a child," but only the right to engage in activities properly ordered toward having one. Infertility can be a profound source of human suffering, but to assert the right to have a child would be to turn human offspring into commodities. Without a prior conception of justice and the moral law, one cannot make sense of these claims made about rights.

Donum Vitae thus uses the personalistic concept of the gratuitous divine gift of human life in order to establish the priority of nature and the natural moral order. The natural law provides a body of concrete moral norms within which liberty is to be exercised. The question that remains is whether the Pope can reconcile his frequent appeal to rights language with his commitment to the defense of the common good. John Paul's later encyclicals *Veritatis Splendor* and *Evangelium Vitae* stand out because they argue forcefully for the absolute dependence of democratic liberties and human rights upon an underlying objective moral order.[21]

5. *CENTESIMUS ANNUS*

George Weigel argues that the three encyclicals: *Centesimus Annus*, *Veritatis Splendor* and *Evangelium Vitae* should be considered part of a single continuous effort to establish "the moral foundations of the free and virtuous society" (Weigel, 2001, p. 757). There are other papal documents that treat this subject, for instance, *Laborem Exercens* and *Sollicitudo Rei Socialis*, but it is the former three that provide the best insight into John Paul's mature position. They all share in common the Pope's personalistic principle, 'the law of the gift', and his companion concern for the responsible exercise of freedom in light of objective moral truth. *Centesimus Annus* observes:

> When man does not recognize in himself and in others the value and grandeur of the human person, he effectively deprives himself of the possibility of benefiting from his humanity and of entering into that relationship of solidarity and communion with others for which God created him. Indeed, it is through the free gift of self that man truly finds himself… A person who is concerned solely or primarily with possessing and enjoying, who is no longer able to control his instincts and passions, or to subordinate them by obedience to the truth, cannot be free: *obedience to the truth* about God and man is the first condition of freedom (John Paul II, 1991, no. 41).

There are significant differences among the encyclicals, most notably the degree to which the priority of the natural moral law features in their arguments. In this respect, *Centesimus Annus* differs markedly from the others. It endorses the priority of moral truth to the genuine exercise of freedom, but emphasizes the personalistic dimension of the Pope's thought. Two features of the document provide a plausible explanation for this emphasis. The first is that the encyclical is pastoral and speculative as opposed to being doctrinal in its approach. The Pope proposes to assess the merits of

Rerum Novarum in light of new developments in history. These developments pose "new requirements of evangelization," however the encyclical is "not meant to pass definitive judgments" (John Paul II, 1991, no.3). The phenomenological method is more suited to this exploration of new and uncertain terrain than a magisterial assessment in light of the principles of the natural law.

Second, John Paul wishes to avoid making categorical assertions about the organization of economic and social life,[22] since that is neither the proper role of the Magisterium, nor his particular competence as a scholar. Given the reality of the fall of Communism, he recognizes that the only serious contender for an economic and political system at present is some form of democracy and the free market. In *Sollicitudo Rei Socialis* he had argued, "...the Church's social doctrine adopts a critical attitude towards both liberal capitalism and Marxist collectivism" (John Paull II, 1987, no. 20).[23] *Centesimus Annus* departs from this fence-sitting posture and embraces democracy as the best way to ensure the dignity of the human person: "The Church values the democratic system inasmuch as it ensures the participation of citizens in making political choices..." (John Paul II, 1992, no. 46). The Pope rightly embraces this economic conclusion in a cautious and pragmatic manner.

Some conservative critics of the Pope perceive a dangerous doctrinal shift in his heightened enthusiasm for liberal democracy. Other personalists, such as Jacques Maritain, went down this road in the past, embracing democracy as the secular fulfillment of a new Christian political ideal.[24] One can see how critics would think that the Pope's personalism goes hand in hand with uncritical acceptance of the modern liberal tradition. Nothing could be further from the truth in this case.

He cautions that forms of capitalism that fail to limit the exercise of freedom by moral and religious principles, are to be rejected. Furthermore, whereas *Rerum Novarum* had treated the right to private property as sacred and inviolable, *Centesimus Annus* argues that this right is "not an absolute value," and that it must be ordered to the common good.[25] Similarly, democracy must be guided by a correct conception of the human person and of "ultimate truth to guide and direct political activity" (John Paul II, 1991, no. 46). When freedom becomes foundationless, or when it becomes the only aim of society, rights themselves are inevitably compromised. The Pope explicitly mentions the right to life in this context, arguing that government support of the violation of this right and others is a clear indication that it has "lost the ability to make decisions aimed at the common good" (John Paul II, 1991, no. 47). The Pope's enthusiasm for democracy, rights and the free market is evidently not captive to modern rights theory, nor does it neglect the priority of an objective moral order, even though the language of natural law is not featured prominently in the text.

The Pope's diagnosis of the events surrounding the fall of communism reinforces this conclusion. He asserts that "An important, even decisive, contribution was made by *the Church's commitment to defend and promote human rights*" (John Paul II, 1991, no. 22). We should not underestimate the influence of external factors upon the collapse of communism, including the West's military and economic engine. The implosion of the various communist regimes could not have taken place so decisively and in many cases bloodlessly, however, without the influence of substantial internal sources of dissatisfaction. The critical flaw in socialism according to the Pope is a distorted sense of the human person and his or her vocation:

> Socialism considers the individual person simply as an element, a molecule within the social organism... Socialism likewise maintains that the good of the individual can be realized without reference to his free choice, to the unique and exclusive responsibility which he exercises in the face of good or evil (John Paul II, 1991, no. 13).

This diagnosis of the fatal weakness in socialism bears the unmistakable stamp of John Paul's philosophical personalism. Consider, for example, the thesis of Jacques Maritain's short book, *The Person and the Common Good.* According to Maritain we must make a fundamental distinction between "personality" and "individuality" (Maritain, 1966). Both communist collectivism and libertarian individualism treat the person as a 'material individual' because they neglect the spiritual dimension of human nature. Material individuals, as distinct from persons, are atomic units of a society. They are deprived of a nature or purpose in the traditional sense and the deeper notion of a spiritual quest that constitutes genuine human fulfillment. Persons, for Maritain on the other hand, have a unique dignity because they are made in the image of God. Personality implies nature, purpose and order to a transcendent good. Furthermore, persons are capable of communion with other human beings and ultimately with God. Such communion is, in fact, the fulfillment of their nature. Persons are inherently social beings. It is not difficult to see how John Paul's 'law of the gift' and the principle of 'solidarity' draw upon the personalist influences that have shaped the Pope's thought.

His diagnosis of the failure of communism is therefore deeply personalistic, yet it is not opposed to the concept of the natural law. Communism fell into difficulty because it had a distorted anthropology:

> Not only is it wrong from the ethical point of view to disregard human nature, which is made for freedom, but in practice it is impossible to do so. Where society is so organized as to reduce arbitrarily or even suppress the sphere in which freedom is legitimately exercised, the result is that the life of society becomes progressively disorganized and goes into decline (John Paul II, 1991, no. 25).

Persons have their dignity because they have a certain kind of nature, and having that nature is the result of a free gift. They are free and responsible beings ordered to a transcendent good. Modern conceptions of individuality neglect this dimension of personality. Personal status provides human beings with certain privileges or rights, but also corresponding responsibilities or duties. Those rights and duties are not arbitrary social constructions. The good of human nature and the moral law that specifies it is prior to any positive legal construction.[26] Critics of the development of the social magisterium have pointed out that the modern language of rights is discontinuous with the priority of law and duty to the exercise of freedom. John Paul's answer to this concern is that a particular conception of human freedom and dignity is at fault. The roots of this conception reach back to the transformation of the notion of 'right' by the late medieval and early modern understanding of freedom as willfulness. For the Thomistic personalist, retrieving the older Thomistic view of 'right' allows for simultaneous use and critique of the language of rights.

Centesimus Annus celebrates the fall of repressive regimes behind the Iron Curtain and the contribution that the Church's defense of human dignity and rights made to that event. It expresses hope for the future of democracy and free markets. There is much beneath the surface of its warm personalistic treatment of human rights and democracy to indicate that the Pope is aware of the need to assert the proper relationship between rights and duties. This proper relationship recalls an older tradition that is unacceptable to many contemporary liberal theorists of democracy and rights. For them it is inconsistent to praise liberty and democratic institutions, while simultaneously tethering them to a substantive conception of the moral good. For John Paul, failure to link the two is precisely why comparable difficulties faced by communism lie in store for contemporary liberal democracies. Both share a distorted conception of the nature of the human person. In their efforts to remove all barriers to the exercise of freedom, including the requirements of an objective moral law, some in the West have lost site of the basis for the moral value of freedom. In doing so they undermine the very defense of freedom and personal dignity they seek to secure. *Veritatis Splendor* and *Evangelium Vitae* shine further light upon this problem, which leads to what the Pope has called "the Culture of Death."

6. *VERITATIS SPLENDOR*

> Today, when many countries have seen the fall of ideologies which bound politics to a totalitarian conception of the world ...there is no less grave a danger that the fundamental rights of the human person will be denied and that the religious yearnings which arise in the heart of every human being will be absorbed once again into politics. This is *the risk of an alliance between democracy and ethical relativism,* which would remove any sure moral reference point from political and social life, and on a deeper level make the acknowledgement of truth impossible... As history demonstrates, a democracy without values easily turns into open or thinly disguised totalitarianism" (John Paul II, 1993, no. 101).

Veritatis Splendor picks up the discussion where *Centesimus Annus* concluded. *Centesimus Annus* focused upon the collapse of socialist regimes, because of their failure to respect the dignity of the human person. *Veritatis Splendor* and *Evangelium Vitae* turn their attention to the moral crises within societies that regard themselves as defenders of the tradition of human rights and liberty. This sort of society constitutes a potentially greater threat to human dignity in the long term, because it is blind to its own deficiencies. Totalitarian regimes deprive citizens of the exercise of their liberties. The "alliance between democracy and relativism" threatens to deprive us of our moral capacity to recognize and defend the dignity of the person. We should not be surprised by the fact that this threat has made its initial appearance in the biomedical arena. A battle necessarily begins at the point where a community's defenses are the weakest. Abortion, euthanasia, physician assisted suicide and various forms of genetic experimentation all constitute offenses against human persons who are least able to assert rights claims. Because civil society exists for the preservation of human rights and promotion of the common good, no civil society can legitimately fail to protect these rights.

Some contemporary liberal theorists, on the other hand, like John Rawls, have argued that a democratic society must be characterized by essential pluralism. Its conception of justice must be procedural, that is, it must abstract from drawing conclusions from any particular substantive vision of the good life. The only substantive good admitted is the exercise of liberty itself. The Pope responds to this challenge in *Veritatis Splendor*, arguing that no free society can be sustained without a shared conception of the moral good. As he remarks in the opening sentence of the encyclical, "truth…shapes" the exercise of freedom. Far from limiting the exercise of freedom, authentic freedom is found in the observation of the requirements of the natural law (1993, nos. 12-13).

For John Paul the best defense of the natural law against the conception of freedom as mere absence of restraint is a correct understanding of the nature of the person. Personalism provides the bridge between the concept of personal dignity and the defense of the natural law. Both socialism and liberal individualism conceive of the person strictly as a subject with physical interests and needs, and of society as a vehicle for the satisfaction of those interests. The Pope's response to this view is that society must be organized so as to protect and nurture a deeper more substantive conception of the human good. Civil society cannot be responsible for producing these goods directly. The principle of subsidiarity suggests that governments are ill suited to the production of other aspects of the common good besides public order and peace. They must protect the right and duty of other groups within society to tend to these needs.

It is in this context of the alliance between democratic institutions and moral relativism that John Paul situates the problems facing contemporary Catholic moral theology. There is a risk that certain "fundamental truths" may be denied because dissent is pervasive and far reaching. The difficulties are manifest, especially in the area of biomedical issues at the borders of human life. He does not mention Paul VI's encyclical *Humanae Vitae* by name, but it is clear that dissent from the Church's teachings on human sexuality has an important role to play in the development of this crisis. In some quarters people no longer regard the Magisterium as capable of teaching genuinely universal moral norms. In place of this view, they argue for an essential separation between the redemptive character of religious faith and the observance of the moral law. So-called "fundamental option" theory, for instance, holds that salvation depends upon a fundamental choice to love God that is deeper than our individual moral choices. In theory, one could reject ordinary morality and remain a good Christian.

The common thread behind these theological innovations, according to John Paul, is ethical relativism born of a distorted sense of human freedom, initiated in large part by the Enlightenment. He recognizes the challenge this conclusion presents, given that he has made the language of rights central to his thinking about the renewal of moral theology. *Veritatis Splendor* makes the case for the papal position by distinguishing between appropriate and inappropriate senses of freedom. It thus differentiates the Pope's view from the liberal rights tradition and provides a response to critics who think John Paul has embraced this tradition too incautiously.

On the positive side he observes that the modern language of rights has contributed to a heightened sense of freedom and the dignity of the human person. This has led to the desire for the enhancement of liberty in many areas, including the exercise of religious freedom, which the Church vigorously supported in Vatican II. The modern concept of rights rests upon some more and less adequate premises though. The first flawed premise is the rejection of our creaturely status as made in God's image. The materialist conception of human nature or rather the very denial of nature alters fundamentally our understanding of the purpose and value of freedom. Freedom is no longer a good that enables us to attain perfect beatitude, but the end itself and "source of all values" (1993, no. 32). Without mentioning Kant directly, John Paul criticizes the Kantian concept of self-legislation. Once we deny the transcendent source of our nature by a divine gift and its teleological ordering to fullness of being, we are not far from relativism, as the purely formal conception of the moral good eventually fails to sustain itself. Perhaps the best evidence for this is the empirical fact of the evolution of moral philosophy in the wake of Kant. Rawls, who was sympathetic to Kant's position, ended up defending the "political conception of justice," which entails relativism about substantive conceptions of the good. The mediating term relating truth and freedom for the natural law tradition is nature. Once nature is rejected, the synthesis inevitably collapses. Nature and freedom are set in opposition to each other, and the objectivity of the moral law loses to the pursuit of absolute liberty.

Veritatis Splendor rejects this false "autonomistic" conception of freedom. Recalling Newman's statement that "conscience has rights because it has duties," the Pope stresses that freedom and the moral law do not contradict each other. This has important implications for his view of the relationship between moral law and civil law. Public authority can serve authentic freedom by promoting the common good consistent with nature, although the principle of subsidiarity limits the scope of such authority theoretically and prudence demands that some public evils are left untouched. The right of religious freedom, for instance, stakes out a realm within society where the government ought not to interfere, even for the sake of the good of truth. The encyclical makes clear that this question can be addressed in the modern idiom with the language of rights, although it is made more perspicuous by the natural law tradition's notion of certain negative moral absolutes specifying moral duties.[27]

Unfortunately, the Enlightenment conception of freedom, standing in opposition to nature, has now penetrated within the theoretical speculation of the tradition itself, placing in jeopardy the correct understanding of the relation between civil authority and the moral law. The central sections of the encyclical give a thorough analysis and critique of these trends in moral theology. A detailed examination of this critique would be tangential to our present purpose. Since the Pope deploys his own constructive view in the course of critiquing these alternative theories, however, we must briefly consider a representative example.

A number of Catholic theologians reject the absolute opposition of freedom and nature, but they try to sustain a measure of free play between autonomy and the objective moral order. The general form of this trend in moral theology is articulated by John Paul as follows:

> ...some authors have proposed a kind of double status of moral truth. Beyond the doctrinal and abstract level, one would have to acknowledge the priority of a certain more concrete existential consideration... [The] circumstances and the situation could legitimately be the basis of certain exceptions to the general rule and thus permit one to do in practice and in good conscience what is qualified as intrinsically evil by the moral law (1993, no. 56).

Although numerous distinctions can be made here, the encyclical places "fundamental option theory" and proportionalism under this heading.[28] According to the concept of the "fundamental option" there is a deeper aspect of our freedom: the fundamental option for or against God, that determines our character as ultimately good or evil. Concrete choices are judged right or wrong on the basis of certain "premoral goods," but our personal goodness or malice is a function of our fundamental option. Persons may make free choices contrary to the moral law that do not alter their fundamental option and their essential goodness.

Related to this view of the fundamental option, we may also speak of proportionalism in moral theology. Proportionalism distinguishes between the good or evil of the agent's intention in making a moral choice and the assessment of the degree of premoral goods promoted or damaged in the process. Because proportionalists separate the order of the goodness of an agent's intention, from the rightness or wrongness of an act based upon the weighing of certain premoral goods, it is impossible to judge a concrete type of moral action to be absolutely prohibited. As John Paul notes, "The moral specificity of acts, that is their goodness or evil, would be determined exclusively by the faithfulness of the person...without this faithfulness necessarily being incompatible with choices contrary to certain particular moral precepts" (1993, no. 75). Without rejecting entirely the idea of transcendent moral precepts, as the materialist and post Kantian conceptions of autonomy do, these positions in moral theology end up in a similar place by isolating the autonomy of conscience from the requirements of an objective moral order.

The Pope responds by stressing two important points. First, following the Thomistic natural law tradition, he asserts that conscience is a judgment of practical reason guided by the natural law. There is therefore no opposition between the proper autonomy of conscientious judgment and the existence of an objective moral order. In fact, "*in the practical judgment of conscience*, which imposes on the person the obligation to perform a given act, *the link between freedom and truth is made manifest*" (1993, no. 61). Second, he appeals to the natural law concept of negative moral precepts that bind absolutely. He argues that it is a mistake to confuse casuistry with the possibility of making exceptions to the observance of the negative moral absolutes, as proportionalists do. Casuistry deals with cases of moral uncertainty, but there has never been any doubt about the certainty of cases governed by the negative moral precepts. Proportionalism tries to resolve all moral choice into a kind of casuistry in which the rightness or wrongness of an act is measured by its propensity to produce certain premoral goods. Against this view, John Paul points out that it is futile to attempt to evaluate all of the consequences of our acts. Even if we could, the consequences of an act are only one set of circumstances that contribute to its moral evaluation, since a set of good consequences cannot make a bad or unjust act choiceworthy. The real moral determinant of an action is the object or proximate end of the agent. Actions governed by the negative moral precepts can never rightly be made the proximate end of the agent, because willing these kinds of actions is inherently disordered.

John Paul uses an example from the realm of sexual ethics in order to illustrate the significance of his point. Proportionalists who dissent from the Church's teachings about contraception tend to treat the issue as a matter for the use of casuistry, in particular as a putative application of the principle of double effect where a lesser evil is tolerated in order to avoid a greater one. The Pope argues, following St. Paul's principle, that it is never permissible to do evil that good may come. In the case of contraception, one acts directly against the good of life for the sake of an ulterior end.

Proportionalism has been used to justify this practice because it rejects the teaching of the negative moral precepts in favor of a more nebulous view of the intention of the agent and the prospect of producing certain premoral goods by an act that is formally contrary to the moral law. The doctrine of the negative moral precepts, on the other hand, stakes out a well-defined subset of moral actions that are universally prohibited because they are always contrary to the natural law.

The negative moral precepts furthermore provide John Paul with the material he needs in order to resolve the question of the relation between civil authority and moral principles, since they identify a range of actions that are always contrary to human dignity. The natural law requires an absolute duty of us never to practice them. To each duty specified by the negative moral precepts, we can enumerate certain inalienable rights that apply to the beneficiary of those duties. Inalienable rights, in turn, indicate the arena in which civil authority must rightly intervene for the sake of the common good.

Outside of the protection of inalienable rights that correspond to the negative moral precepts of the natural law, it is not appropriate for governments to try to produce full virtue in their citizens, because that is beyond their role and competence. Hence, the right of religious freedom must be defended, even in cases of erroneous conscience, not because an erroneous conscience is good, but for the sake of the freedom to worship God in truth. The violation of the negative moral absolutes, however, constitutes a breach of the common good that cannot be tolerated and which civil authority must rightly address. Merely procedural justice is not enough. Liberal individualist democracy and totalitarianism share the same dangerous potential to eclipse moral truth in favor of a distorted understanding of authentic freedom and the common good. True to his efforts to present his natural law argument in terms intelligible to the modern sense of subjectivity, John Paul bolsters it with a remarkable personalist observation. Citing the example of the martyrs, he points out that denial of moral absolutes trivializes their witness to the reality of an objective moral order, just as their sacrifice testifies more forcefully to its reality.

7. *EVANGELIUM VITAE*

> This is what is happening also at the level of politics and government: the original and inalienable right to life is questioned or denied... This is the sinister result of a relativism which reigns unopposed... In this way democracy, contradicting its own principles, effectively moves towards a form of totalitarianism (1995, no. 20).

What was a worry in *Centesimus Annus*, and a concern about the direction of moral reasoning in *Veritatis Splendor*, has become a full blown social and political crisis in *Evangelium Vitae.* It would appear that cautious optimism about democratic institutions and the rights tradition has given way to profound disappointment and regret. Speaking of the lack of will on the part of international institutions such as the U.N. to protect basic human rights, the Pope argues that a failure to address the crisis will show the world's previous statements to have been a mere rhetorical exercise. The attack upon life represents, "a direct threat to the entire culture of human rights...jeopardizing the meaning of democratic coexistence" (1995, no. 70).

We see in *Evangelium Vitae* a return to the use of the language of rights, but profound concern about where the use of that language is heading. Foremost among the developments that concern the Pope is the tendency within liberal democracies to reject a relationship between civil authority and a substantive conception of the moral good, due to the growth of ethical relativism. John Paul reiterates John XXIII's statement from *Pacem in Terris* that the common good is best served when we protect individual rights, but he cautions that we must not see democracy as an end in itself, or as a substitute for the observance of moral absolutes.

Attention to the audience and pastoral concerns of the document helps the reader to place its particular emphasis in perspective. *Veritatis Splendor* was addressed to the bishops, hence it makes freer use of the terminology of the natural law tradition, with which the Episcopal audience would be familiar. *Evangelium Vitae* is addressed to "all people of good will." The language of rights thus has a more prominent place, although the language of natural law and duty continues to be used. *Veritatis Splendor* dealt primarily with abstract principles of moral reasoning, whereas *Evangelium Vitae* focuses upon concrete instances of moral choice. The target audience is clearly people and governments who profess strenuously to be proponents of human rights, such as Western liberal democracies. Hence, *Evangelium Vitae* deals most directly and extensively with biomedical issues, which the Pope sees as the final and most dangerous of the 20th century's long array of attacks on the human person. Not surprisingly, it also contains John Paul's most complete statement of the relation between civil authority's need to respect fundamental rights and to defend substantive moral principles.

The Pope seeks to enter into dialogue with an audience that is immersed in the tradition of rights. A careful reading of the text demonstrates that he embraces the modern idiom as he simultaneously attempts to transform it. He begins with a distinctively modern and personalist consideration, the inherent dignity of the person, but he grounds this dignity in his 'law of the gift'. The dignity of the person comes not from an inherent possession, but from a divine transcendent gift. As we have already observed, that source of dignity opens up the possibility of a very different sense of personal autonomy and freedom, since the extrinsic source of the gift implies many responsibilities. Personalism, as opposed to abstract theoretical principles, is the vehicle for the argument, although it is designed to place subjectivity in harmony with the objective order of the natural law.

Having established the basis for his interpretation of human dignity, he turns to the crisis that is his primary concern: an unprecedented attack upon the basic good of life is being mounted in the name of rights and individual freedom. Although this development is surprising to John Paul because of the progress that has been made towards the protection of human rights in other areas, he is certainly aware of its causes and implications. As he had previously argued in greater detail in *Veritatis Splendor*, he reiterates that the basic cause of the difficulty is an exaggerated sense of freedom, deriving from the Enlightenment conception of the individual. Autonomy on this view equals absolute freedom to dispose of things as one sees fit, because the agent's freedom is an absolute end-in-itself.

In contrast to the Enlightenment view, the Pope stresses solidarity or the "inherently relational dimension" of our interaction with others, as well as genuine freedom's essential relationship to moral truth. Once we isolate ourselves from all forms of authority, the value of freedom itself can be called into question, if not for ourselves, for other persons who are less capable of defending themselves. This exaggerated sense of autonomy is part of a "caricature" of authentic democracy, which originally acknowledged its dependence upon the existence of an objective moral order. He postulates two sources, one proximate and the other ultimate for this decline: the alliance of democracy with relativism, and the "eclipse of the sense of God and of man," both of which he had identified in *Veritatis Splendor*. The crisis of Western liberal democracies is ultimately a religious one: a loss of faith. This is a vitally important observation, since the Pope had previously endorsed democracy as the best sort of regime to protect individual rights, including the right to religious freedom. Yet, he now decries secularism as the driving force behind relativism and the eclipse of god. If as some think, relativism is an essential condition of democracy, then the Pope is deeply confused. His response to this apparent predicament is that our understanding of democracy has undergone an unhappy and unnecessary transformation to the sort of political system envisaged by Rawls and some other liberal political theorists.

According to the contemporary view autonomy is absolute. Each one seeks his or her own well being, but in the face of competing interests a compromise must be found. The purpose of the compromise is to provide maximal concomitant liberty for each individual. We assume that a shared common view of objective moral truth is impossible because it is presumed that freedom of conscience will result in irreducible pluralism. Once irreducible pluralism is assumed, the state must confine itself to legislating only those norms that fall within the consensus of the majority of citizens. As Rawls argued, such an agreement would need to be procedural rather than substantive. We may agree about certain structures of fairness, but a substantive conception of the moral good is out of the question. The political conception of justice will necessarily privatize and trivialize our substantive conception of the good. The Pope reports of this view that:

> …in the exercise of public and professional duties, respect for other people's freedom of choice requires that each one should set aside his or her convictions in order to satisfy every demand of the citizens which is recognized and guaranteed by law… Individual responsibility is thus turned over to the civil law, with a renouncing of personal conscience, at least in the public sphere (1995, no. 69).

In recent years, the liberal individualist position has provided a convenient justification for the position of many American Catholic politicians on critical biomedical questions. They maintain that they are personally opposed to the morality of abortion, euthanasia and genetic experimentation, but that respect for the rights of others who do not share their substantive moral convictions requires them to support public measures contrary to those convictions. There is little if any connection between the exercise of civil authority and a substantive conception of the moral good. The Pope's response to this line of argument is frank and direct: "…what we have here is only the tragic caricature of legality; the democratic ideal which is only truly such when it acknowledges and safeguards the dignity of every human person, *is betrayed in its very foundations*…"(1995, no. 20). In short, the Pope asserts the necessary complementarity between the exercise of civil authority and the safeguarding of objective moral principles.

Contrary to the modern tendency to regard democracy as an end in itself, *Evangelium Vitae* stresses that it is one of a number of political systems, whose value is to be measured by the ends it pursues and by its capacity to remain in conformity with moral truth. Consistent with the argument of *Centesimus Annus* though, John Paul reaffirms the view that democracy is best suited to the protection of the dignity of the person. At this point, the reader is bound to wonder whether it is possible to hold together this optimistic appraisal of democratic institutions with profound concern about its tendency towards moral relativism.

John Paul's answer to this concern is that there is no insurmountable difficulty with democratic polities and the language of rights when both are grounded in a prior objective moral order. This order, specified by the principles of the natural law, is the necessary basis for civil society. *Evangelium Vitae* underscores the point that the papal tradition of appealing to the language of rights is committed to an older more inclusive notion of the relationship between rights and duties. John Paul quotes John XXIII, who was in turn quoting Pius XII:

> The chief concern of civil authorities must therefore be to ensure that these rights are recognized…and that each individual is enabled to perform his duties more easily. For "to safeguard the inviolable rights of the human person, and to facilitate the performance of his duties, is the principal duty of every public authority" (1995, no. 30).

Marxist and liberal individualist conceptions of the person share the same flawed presuppositions. Persons are not regarded as having a unique dignity that establishes for them a teleological order to human fulfillment. This promotes a false opposition between freedom and the moral law, because it denies the ordination of nature to its proper end. For the Rawlsian, the self is radically 'unencumbered'. Autonomy and freedom are thus conceptualized as contrary to the requirements of an objective moral order. For John Paul, however, freedom cannot be understood without reference to the objective moral order, which is none other than the rational specification of the human good.

Because freedom and the moral order are complementary, civil authority cannot neglect a substantive conception of the human good. The difficulty is to establish the appropriate limits of civil authority, once this point has been granted. The right to religious freedom, for instance, depends upon the idea that the promotion of some of the virtues is beyond the appropriate competence of the state. *Evangelium Vitae* addresses this issue, noting that: "*the purpose of civil law* is different and more limited in scope than that of the moral law" (1995, no. 71). Here John Paul draws upon the scriptural and Augustinian notion of the 'tranquility of order' as a means to establish the appropriate limits for civil authority. The essential purpose of the state is to provide the conditions for peace and justice for its citizens. It must tolerate certain evils for the sake of the common good, since the effort to eradicate them would be futile or even produce greater evil. Not withstanding these limitations of civil authority, the state must protect certain inviolable human rights, including the right to life, without which the state itself becomes radically disordered.

John Paul reiterates, what he had argued much more extensively in *Veritatis Splendor*, that these fundamental rights correspond to a class of actions (or injuries) that are prohibited by the negative moral absolutes specified by the divine and natural laws. Thus, rights and duties are complementary. We have a duty to avoid doing especially those actions that violate the negative moral precepts, because such actions are wrong in kind (*per se malum*). No choice to violate them can ever be judged correct. Because of their special status as universally impermissible, no person can therefore legitimately claim a right to do such actions, nor can a state claim the right to permit the dignity of persons to be violated by such injuries. John Paul thus demonstrates how the language of inalienable rights can be assimilated to the natural law terminology of moral absolutes. Contemporary liberal individualist views, on the other hand, abstract from substantive truth claims in favor of procedural ones concerning the maintenance of fair and equal political processes. As the Pope argues, all such views must eventually collapse into relativism. Without a conception of the objective order of the natural law, democratic political institutions have been "reduced to a mere mechanism for regulating the different and opposing interests on a purely empirical basis" (1995, no. 70). The urgent appeal of *Evangelium Vitae* is that we should "rediscover those essential and innate human and moral values which flow from the very truth of the human being and express and safeguard the dignity of the person" (1995, no. 71). John Paul is certainly aware of the difficulty of achieving this hope, since he argues that it will require the reverse of the tide of the 'eclipse of God' in modern culture.

Concretely, his defense of the necessary but limited relation between moral principles and the exercise of civil authority has significant implications for Western democracies and their citizens. The negative moral absolute prohibiting killing corresponds to an inviolable right to life for all innocent persons. All forms of threats against life, including abortion, euthanasia, assisted suicide, destructive research with human embryos and the like cannot justly be made legal by civil authority. Because of the gravity of the threats against life for the foundation of the political order, individual citizens and legislators have a "*grave and clear obligation to oppose them by conscientious objection*" (1995, no. 73).

Here it is worth noting that politicians who support legislation designed to diminish threats against life incrementally, when no better alternative remains open, do not constitute an exception to the principle previously articulated. The Pope considers the example of legislation, such as that prohibiting partial birth abortion, when a more restrictive law is presently impossible to pass. In that circumstance, he argues, one can legitimately vote for the law because, "This does not represent an illicit cooperation with an unjust law, but rather a legitimate and proper attempt to limit its evil aspects" (1995, no. 73). Illicit cooperation here is "formal cooperation in evil" where one commits to "direct participation in an act against innocent human life or a sharing in the immoral intention of the person committing it' (1995, no. 74). This is opposed to material cooperation where one legitimately acts for the sake a good, foreseeing that other goods will as a consequence be damaged, but the damage to those goods is no part of the agent's plan. It is merely accepted as an inevitable though undesired concomitant consequence or side effect of acting for the sake of another good.

Contrary to the proportionalist view discussed above, this is a legitimate application of the traditional principle of double effect, not a reversal of the negative moral precepts. It is beyond the scope of the present inquiry to offer a detailed analysis of this point, but we may note that *Evangelium Vitae* depends at this point upon the more extensive discussion of the moral determinants of an action in *Veritatis Splendor*. As John Paul argued in the previous encyclical, the moral quality of an action is determined in the first instance by the object or proximate end of the agent's action, and not by circumstances such as its consequences. A legislator who publicly opposes abortion and votes for partial birth abortion legislation has as his or her proximate end the limitation of the practice of abortion. Other abortions will continue to take place; they are foreseen though unintended consequences of the action. They are neither a means to the proximate end of limiting certain types of abortion, nor an end in their own right. The principle of the necessary relation between civil authority and objective moral norms therefore shows itself to be flexible enough to accommodate certain demands of political prudence, without collapsing into the moral relativism of competing views.

8. CONCLUSIONS

The field of biomedical ethics covers a very broad range. Recent papal teaching on the subject has been organized around a single coherent theme often repeated in *Evangelium Vitae*: the Church must respond to what John Paul has labeled "The Culture of Death." Abortion, euthanasia, physician-assisted suicide, cloning and embryonic stem-cell research, among other practices, involve attacks upon the sanctity and integrity of human life. One can see that this concern runs much deeper in the Pope's thinking. From biomedical ethics to geopolitical events such as famine, genocide and the just distribution of the world's economic resources, the common thread tying together his social teaching is a concern for the dignity of the human person as a creature made in God's image.

The "Culture of Death," as the Pope sees it, is a betrayal of that for which modernity, including especially the rise of representative democratic government, is rightly to be praised. It is a direct outgrowth of theoretical approaches such as the ones we have just seen criticized. The primary idiom that modern authors and statespersons have used in order to fight for increased respect for human persons is the language of rights, but it is unfortunately the "culture of rights" that John Paul sees as spinning out of control, since it has lost touch with moral truth and Gospel teachings about the human person:

> …with tragic consequences, a long historical process is reaching a turning-point. The process which once led to discovering the idea of "human rights" – rights inherent in every person and prior to any Constitution and State legislation – is today marked by a *surprising contradiction.* Precisely in an age when the inviolable rights of the person are solemnly proclaimed and the value of life is publicly affirmed, the very right to life is being denied or trampled upon, especially at the more significant moments of existence: the moment of birth and the moment of death (1995, no. 18).

The Pope's attempt to dialogue with the culture of rights while critiquing it has led to some difficulties. Whereas liberal political theorists dislike his application of the dignity principle to the unborn, conservative theorists charge that the present form of the dignity principle constitutes an unhealthy intrusion of Kant and the Enlightenment into Christian theology, one which inevitably leads to moral subjectivism.[29] The preceding analysis shows, however, that John Paul has not embraced the culture of rights uncritically in an attempt to achieve currency for the Church's social teaching; rather, he seeks to engage liberal thought by acknowledging its strengths, while simultaneously reminding it of the requirements of an objective moral order.

Personalism, appropriately tempered in the Pope's case by a commitment to Thomistic moderate realism, allows him to speak the language of modern subjectivity without succumbing to its relativistic and radical individualistic excesses.

Far from baptizing democratic forms of life as the embodiment of Christian political aspirations, the Pope asserts that the "Church has no models to present..."(1993, no. 43). He argues that liberal democracy is just as liable to totalitarian excesses as Marxism and socialism.[30] These excesses emerge when democracies deny the existence of transcendent truth and the objective moral order, but he refuses to accept that state sponsored atheism and moral relativism are essential to democracy.

Modern rights theorists do tend to pit the rights of the individual against duty, the moral law and the common good. There is an obvious sense in which the language of 'right(s)' has been transformed by the atomistic conception of the individual and the rejection of notions such as teleology, nature and the priority of the common good. These terminological transformations have important implications for the field of bioethics. The question at issue is whether these obvious and important differences in terminology create a gap that cannot be bridged between the language of natural law and that of natural rights in such a way as to preserve the tradition while engaging the modern world in its own terms. In other words, is there a way of speaking the Kantian language of the dignity of the person without failing to recognize the priority of duty, law and nature? John Paul is certainly aware of this need. Consider the case of Karol Wojtyla's earlier book on sexual morality, *Love and Responsibility.* This work is sympathetic to Kant's formulation of the 'Love Commandment' as the categorical imperative requirement to treat persons as ends in themselves and never merely as means to an end (Wojtyla, 1993d, pp. 40-44). Yet, it offers a criticism of Kant's conception of autonomy and defends the priority of the objective moral order.

In the final analysis, the best elements of John Paul's papal writings do succeed in formulating a way of embracing contemporary rights theory while preserving important elements of the natural law tradition. Moreover, the search for this middle path is a longstanding concern of the Pope. Liberal critics of John Paul perceive him to be a reactionary. Conservatives worry on the contrary that he has succumbed to modern liberal subjectivism. The core of his philosophical and theological teaching is neither liberal nor conservative, but Christocentric (John Paul II, 1997, no. 2). His constant focus upon the divine gift of human life and the gratuitous act of love that is its source appears to have had the salutary benefit of providing him with a means of connecting the natural law tradition's concern with objective moral truths and the common good, to the contemporary concern with subjectivity and the language of rights. This allows him to be consistently both an admirer and a critic of contemporary liberal democracy.

Assumption College
Worcester, Massachusetts

NOTES

1 See e.g. John Paul II, 1984, no. 44; 1991, no. 46.

2 For a detailed discussion of Bishop (later Archbishop) Karol Wojtyla's contributions to the documents of the Second Vatican Council, including his substantial contributions to *Gaudium et Spes* and *Dignitatis Humanae*, see Weigel, 1991, pp. 145-180.

3 For his statement on evolutionary theory see John Paul II, "Message to the Pontifical Academy of Sciences," October 22, 1996. See also e.g. John Paul II, "*Message to Rev. George V. Coyne, S.J.*," in Robert J. Russell, William R. Stoeger, S.J., & George V. Coyne, S.J., eds., 1988.

4 See Wojityla "On the Metaphysical and Phenomenological Basis of the Moral Norm in The Philosophy of Thomas Aquinas and Max Scheler," in Sandok, trans., 1993.

5 See Wojtyla, "Participation or Alientation," in Sandok, trans., 1993. See also Wojytla, 1979, p. 300ff.

6 For the use of this term to describe John Paul's basic intellectual stance, see Weigel, 1999, pp. 136-137.

7 This theme occupies a prominent place in John Paul's encyclicals *Veritatis Splendor* and *Evangelium Vitae*. For an early philosophical expression of the importance of self-transcendence, see Wojtyla, 1993b. See also Wojtyla, 1979. The Christocentric dimension of neighbor love is treated in *Sign of Contradiction*.

8 See *Love and Responsibility*, p. 247.

9 For a discussion of these points, see Karol Wojtyla, 1993c; also Wojtyla, 1979.

10 See e.g. Wojtyla, 1993a.

11 See *Dignitatis Humanae* no. 3. In fact, *DH* commonly links the right of religious freedom with duty. No. 1 speaks of "responsible freedom" animated by a "sense of duty." No. 6 speaks of the principal duty of civil society to protect the common good as being composed not only of the "protection of rights," but also the protection for the "performance of the duties of the human person."

12 See e.g. Kraynack, 2001; also Fortin, 1997.

13 See e.g. John Paul II, 1995, nos. 69-70

14 The document is signed by Joseph Cardinal Ratzinger as the Prefect and Alberto Bovone, Titular Archbishop of Caesarea in Numidia as Secretary. The combination of these signatures indicates the solemnity and importance of the publication. The fact that *Donum Vitae* was promulgated on the Feast of the Chair of St. Peter, would suggest the Pope's own personal involvement with its content and message.

15 See e.g. *Donum Vitae*, Introduction, no. 4.

16 See e.g. the discussion of the "rights" of spouses in nos. II.2, 4, 7, 8 and III.

17 See *Donum Vitae*, Introduction, no. 1.

18 See also *Donum Vitae*, Introduction, no. 5; nos. I.2, 6; III.

19 This precise formulation of the right to become a parent solely through one's spouse cannot be incidental, because it is repeated verbatim in the *Catechism of the Catholic Church* (at no. 2376) and a very similar remark is made in *Evangelium Vitae* (at no .14). It also recalls the teaching of John Paul's predecessor Paul VI in *Humanae Vitae*.

20 See e.g. *Donum Vitae* no. II.4, "Contraception deliberately deprives the conjugal act of its openness to procreation and in this way brings about a voluntary dissociation of the ends of marriage… The moral relevance of the link between the meanings of the conjugal act and between the goods of marriage, as well as the unity of the human being and the dignity of his origin, demand that procreation of a human person be brought about as the fruit of the conjugal act specific to the love between spouses."

21 See also e.g., John Paul II, 1993, nos. 87-89, and John Paul II, 1995 nos. 19-20.

22 John Paul II, 1991, no. 43 and Weigel, 1999, p. 613.

23 For a discussion of this point see Weigel, 1999, pp. 559-560.

24 See e.g. Maritain, 1968, pp. 162 ff.

25 See Leo XIII, no. 46; John Paul II, 1991, nos. 6, 30.

26 This offers a plausible way of interpreting the Pope's apparently un-Thomistic assertion that the person is in some sense prior to the state. See e.g. John Paul II, 1991, no. 13.

27 For a useful discussion of this point in line with the basic argument of the encyclical see Finnis, 1998.
28 This is discussed in no. 65 and proportionalism is discussed in nos. 74-75.
29 See e.g, Kraynack, 2001 pp. 109, 152-159. See also Fortin, 1987, pp. 19-22, 49-56, 191-229.
30 John Paul II, 1991, no. 44. See Maritain, 1966, p. 91.

REFERENCES

Congregation for the Doctrine of the Faith. (1988). *Donum Vitae*. Available at http://www.vatican.va/roman_curia/congregations/cfaith/documents/rc_con_cfaith_doc_19870222_respect-for-human-life_en.html.

Fortin, E. (1997). *Human Rights, Virtue and the Common Good: Untimely Meditations on Religion and Politics*. B. Benestad, (Ed.), Lanham, Md: Rowman and Littlefield.

Finnis, J. (1998). *Aquinas: Moral, Political, and Legal Theory*. New York and Oxford: Oxford University Press.

John Paul II (1979). *Redemptor Hominis*. In J.M. Miller, CSB. (Ed.), (2001). *The Encyclicals of John Paul II*, 2nd Edition. Huntington, Indiana: Our Sunday Visitor Press.

John Paul II (1984) *Sollicitudo Rei Sociali*. in In J.M. Miller, CSB. (Ed.), (2001). *The Encyclicals of John Paul II*, 2nd Edition. Huntington, Indiana: Our Sunday Visitor Press.

John Paul II (1988). "*Message to Rev. George V. Coyne, S.J.*," in R.J. Russell, W.R. Stoeger, S.J., and G.V. Coyne (Eds.), *Physics, Philosophy, and Theology: A Common Quest for Understanding*, South Bend: University of Notre Dame Press, pp. 1-14.

John Paul II (1991). *Centesimus Annus*. In J.M. Miller, CSB. (Ed.), (2001). *The Encyclicals of John Paul II*, 2nd Edition. Huntington, Indiana: Our Sunday Visitor Press.

John Paul II. (1993). *Veritatis Splendor*. In J.M. Miller, CSB. (Ed.), (2001). *The Encyclicals of John Paul II*, 2nd Edition. Huntington, Indiana: Our Sunday Visitor Press.

John Paul II. (1995). *Evangelium Vitae*. In J.M. Miller, CSB. (Ed.), (2001). *The Encyclicals of John Paul II*, 2nd Edition. Huntington, Indiana: Our Sunday Visitor Press.

John Paul II (1996). "Message to the Pontifical Academy of Sciences." Available at http://www.cin.org/users/james/files/message.htm.

John Paul II (1998). *Fides et Ratio*. In J.M. Miller, CSB. (Ed.), (2001). *The Encyclicals of John Paul II*, 2nd Edition. Huntington, Indiana: Our Sunday Visitor Press.

Kraynack, R. (2001). *Christian Faith and Modern Democracy: God and Politics in the Fallen World*. South Bend: University of Notre Dame Press.

Leo XIII (1890). *Rerum Novarum*. Available at http://www.vatican.va/holy_father/leo_xiii/encyclicals/documents/hf_l-xiii_enc_15051891_rerum-novarum_en.html.

Maritain, J. (1966). *The Person and the Common Good*. John J. Fitzgerald, trans., South Bend: University of Notre Dame Press.

Maritain, J. (1968). *Integral Humanism: Temporal and Spiritual Problems of a New Christendom*. New York: Charles Scribner's Sons.

Miller, J.M., CSB. (Ed.), (2001). *The Encyclicals of John Paul II*, 2nd Edition. Huntington, Indiana: Our Sunday Visitor Press.

Weigel, G. (2001). *Witness to Hope: The Biography of John Paul II*. New York: Cliff Street Books / Harper Collins.

Wojytla, K. (1979). "The Person: Subject and Community," *Review of Metaphysics*, XXXIII:2, pp. 273-308.

Wojtyla, K. (1993a). "On the Metaphysical and Phenomenological Basis of the Moral Norm in The Philosophy of Thomas Aquinas and Max Scheler," in Theresa Sandok, trans., *Person and Community: Selected Essays*, New York: Peter Lang, pp. 73-94.

Wojtyla, K. (1993b). "Participation or Alienation," in Theresa Sandok, trans., *Person and Community: Selected Essays*, New York: Peter Lang, pp. 197-207.

Wojtyla, K. (1993c). "Thomistic Personalism," in Theresa Sandok, trans., *Person & Community: Selected Essays*, New York: Peter Lang, pp. 165-175.

Wojtyla, K. (1993d). *Love and Responsibility*, H.T. Willetts trans., San Francisco; Ignatius Press.

Second Vatican Council. (1975). *Declaration on Religious Freedom, Dignitatis Humanae*. Available at http://www.vatican.va/archive/hist_councils/ii_vatican_council/documents/vat-ii_decl_196551207_dignitatis_humanae_en.html

CHAPTER FIVE

MARK J. CHERRY

BIOETHICS IN THE RUINS OF CHRISTENDOM: WHY JOHN PAUL II'S DIAGNOSIS REQUIRES A MORE RADICAL CURE THAN MAY AND COLVERT PROVIDE

1. INTRODUCTION

William May's and Gavin Colvert's and rich and insightful essays on the moral theology of John Paul II critically diagnose a profound shift in moral commitments within the dominant intellectual culture of western Europe and the United States. As each notes, these changes have been especially prominent in medicine and medical morality. Where abortion had once been forbidden, it is now widely practiced. Where the destruction of human embryos had once been understood as equivalent to murder, it has become more or less routine. Third-party assisted reproduction, such as in vitro fertilization, frequently results in both embryo wastage and elective abortion for fetal reduction. Such practices enjoy legal protection as part of a tradition of procreative liberty and for all practical purposes no longer appear outside of the social norm.[1] Moreover, there exists the expectation that significant medical developments will follow from basic research on human embryos. The dominant secular bioethics simply fails to comprehend the destruction of embryos or fetuses as homicide.[2] Indeed, as May and Colvert are aware, the bioethics dominant in the United States and western Europe frequently lays claim to a universal account of proper moral deportment, including the foundations of law and public policy to guide healthcare decisions and medical research, divorced from traditional Christian moral and religious commitments.

In the face of such challenges, how can one sustain traditional Christian medical morality? How can one sustain a culture that supports the cardinal elements of human flourishing? As May and Colvert are aware, answers to these fundamental and pressing questions depend on one's background moral anthropology. Each notes that John Paul II frames the justification for biomedical practices in Roman Catholic anthropology, which, they argue, is grounded firmly in the natural law. John Paul II thereby sustains, they urge, thick concepts of the common good and human dignity, together with the basic goods and rights that constitute the central elements of moral practical reason. Here, however, conflicts among moral anthropologies should be appreciated as especially salient: different background moral anthropologies will ground divergent accounts of human flourishing, and

C. Tollefsen (Ed.), John Paul II's Contribution to Catholic Bioethics, pp. 73–92.

thereby sustain diverse moralities. Thus, as this paper will explore, one cannot make sense of human nature, the ordering of human goods, and the cardinal elements of human flourishing without the appropriate theological understanding of man's relationship to the Creator; in particular, of the relationship among human goods and the demands of God. Critically assessing the norms which ought to guide medicine and bioethics cannot be undertaken merely as an intellectual or political challenge, nor can particular bioethical concerns, such as abortion and embryocide, be adequately addressed through secular reason. Each much be understood more fundamentally in hierological terms.

From a traditional Christian point of view, morality, and thereby bioethics, is best understood within categories that transcend the right, the good, the just, and the virtuous; namely, the holy. Reliance on discursive human reason will at best provide a one-sided incomplete approach to the challenges of addressing the dominant secular morality of our contemporary culture. Adequate response to abortion and embryocide, as well as to the pope's diagnosis of the culture of death more broadly construed, depends on fundamental philosophical and theological issues, including the character of an appropriate philosophically and theologically anchored anthropology, where the central element of traditional Christian anthropology is that humans are created to worship God. Christian morality and moral epistemology must be nested within and understood through this background Christian anthropology. If humans are created to worship God, their morality must be supportive of and incidental to this cardinal, ontological characteristic. One cannot make sense of human nature, the proper ordering of basic human goods, and the cardinal elements of human flourishing without intimate interaction with the Creator. Worshiping rightly is central for coming to know truly. In short, the moral life depends on understanding human good in terms of Divine reality in a significantly more encompassing sense than either May or Colvert have suggested. If humans are created beings intrinsically oriented towards worship, then failures to worship rightly will ineluctably lead to misperceptions of morality in general and bioethics in particular.

As a result, I will argue in the following section that rational philosophical appeal to the natural law is not up to the task of satisfactorily addressing the culture of death, including its implications for medicine, abortion and embryo research. The pope's diagnosis requires a more radical cure: Christian medical morality and bioethics depend on a more fundamental spiritual-therapeutic approach. This approach will be developed in section III with respect to the proper role of medicine in a traditional Christian life, in section IV with regard to the sources authorizing knowledge about that proper role, and in section V with respect to conclusions concerning the reshaping of medical morality and bioethics. In fine, the moral life depends on understanding human goods in terms of Divine reality. Medicine's ability to treat somatic and psychological illnesses can only be appropriately appreciated when that use is placed within the larger framework of curing the soul.

2. THE NATURAL LAW AND COMPETING VISIONS OF HUMAN FLOURISHING

A. Natural Law, Morality, and Public Policy

Natural law moral philosophy embraces particular metaphysical and epistemological foundations. It seeks to address permissible individual and political choice through an objective understanding of human nature and the human good. As a result, natural law seeks principles and precepts for morality, law, and other forms of social authority, whose prescriptive force is not dependent for validity on human decision, social influence, past tradition, or cultural convention (Boyle, 1999); as a field of inquiry, natural law morality understands itself as truly universal. Moreover, these universal moral norms are held to be knowable through human practical reason. Thomas Aquinas held that the natural law is a function of reason, "...promulgated by the very fact that God instilled it into man's mind so as to be known by him naturally" (ST, I-II, Q90, A4). As Colvert argues, the natural law "expresses and lays down the purposes, rights and duties which are based upon the bodily and spiritual nature of the human person" (Colvert, 2004, p. 55, quoting from *Donum Vitae* no. 3). Moreover, it "...provides a body of concrete moral norms within which liberty is to be exercised" (Colvert, 2004, p. 56). According to natural law morality, persons are obligated to will and to act in ways that are compatible with creating and integrating the basic human goods into one's life and the lives of others. Such goods provide the basis for reasoning about virtuous choices; that is, attainment of the basic human goods provides immediate reasons for action.[3]

Similarly, May argues that natural law moral norms should be appreciated as facilitating the attainment of the human good or telos – the goal of full human flourishing. It is the intentional shaping of one's character so that one instantiates and desires those goods central to human fulfillment.[4] May states: "...the core of the action is the free, self-determining choice that abides in the person, making him or her to be the kind of person he or she is" (May, 2004, p. MS 37). The goal is not a maximization of such goods, but the making of rational choices in the pursuit of a virtuous, flourishing life.[5] Not all ends may be permissibly chosen since many goals are incompatible with the basic human goods. Certain actions, such as intentionally killing the innocent, are prohibited since directly willing to kill the innocent is held never to be compatible with human flourishing (see Boyle, 2004). "Intrinsically evil acts violate ... and 'radically contradict' ...'the good of the person, at the level of the many different goods which characterize his identity as a spiritual and bodily being in relationship with God, with his neighbor, and with the material world ..." (May, 2004, p. 40). One respects persons, May argues, by respecting the goods central to human fulfillment. Such norms should be embodied in the lives of virtuous individuals in such a way that what they desire and what is morally required are identical (see, e.g., Boyle, 1999, p. 112).

As noted, natural law moral norms are believed to be more than religiously inspired external constraints on choice and action, but rather to provide the content and requirements of practical reasoning. Thus, as May urges, the natural law directly informs permissible social institutions and public law: "The truth of these moral absolutes is rooted in the primordial principle of natural law requiring us to love our neighbors – beings who, like ourselves, are persons made in the image of

God and who, consequently, have an inviolable dignity. …these norms represent the unshakable foundation and solid guarantee of a just and peaceful human co-existence, and hence of genuine democracy…" (May, 2004, p. MS 42). Colvert argues similarly that there is a necessary connection between the natural law's account of objective human goods and the legitimate function of civil authority: "Foremost among the developments that concern the Pope is the tendency within liberal democracies to reject a relationship between civil authority and a substantive conception of the moral good, due to the growth of moral relativism" (Colvert, 2004, pp. 63-64). For example, in the context of abortion and in vitro fertilization, Colvert notes that "*Donum Vitae* argues that civil law has no authority to grant the permissibility of the use of reproductive techniques that distort or remove the rights inherent to spouses" (Colvert, 2004, p. 59). In short, even where the pope appears to endorse particular, even at times idiosyncratic, Roman Catholic moral claims, Colvert and May argue that such norms, including the limits of permissible individual choice and social policy, are objectively grounded in the natural law and are, in principle, knowable through human reason.[6]

B. Competing Moral Visions of Human Flourishing

This fundamental commitment to human reason within May's and Colvert's Christian moral analysis is somewhat puzzling. On the one hand, it tends to reduce moral theology and Christian faith to secular philosophical analysis; and, on the other hand, it retains what appear to be somewhat accidental connections to the particularities of Christian morality. Such an epistemology commits one to the position that Christians do not have special moral knowledge through correct faith in and right worship of God and, moreover, that Roman Catholics should be able to convince all persons of the truth of Christian moral norms through secular reason, independent of a confession of faith. Yet empirically, reliance on secular reason has failed to unify all of humanity in a common content-full Roman Catholic moral vision. Instead of unity, one finds a considerable pluralism of contradictory and non-reducible religious and secular accounts of the goods central to human flourishing, as well as significantly diverse theories for rationally debating the merits of these substantially divergent understandings of human nature. There are as many different moral anthropologies as there are major world religions and competing secular worldviews.[7]

Consider as heuristic, abortion and embryocide: May and Colvert each agree that abortion, the destruction of embryos, and the use of embryos for research purposes are intrinsically evil. May argues:

> From this one can easily conclude that the anthropology and moral philosophy/theology rooted therein which we find in *Veritatis splendor* holds as utterly immoral the choice intentionally to kill innocent human beings, no matter what their stage of development or the quality of their lives. It thus condemns, … the following: abortion chosen as either means or end (=direct or intentional abortion) (cf. *EvangeliumVitae,* nos. 58-62) …, using human embryos as "research material" or as providers of organs or tissues for transplants to other persons (cf. ibid., no. 63), the killing of human embryos to obtain their stem cells for research and/or therapeutic use on *other* human subjects (May, 2004, p. 47).

Colvert similarly concludes that "abortion... and embryonic stem-cell research, among other practices, involve attacks on the sanctity and integrity of human life" (Colvert, 2004, p. 68). Yet, such a position hardly follows in an obvious manner from general secular practical reasoning (Lustig, 2004).

The dominant secular moral culture endorses abortion as essential for the liberation of women from the blind forces of reproductive biology and as necessary to preserve gender equality, removing the evils of patriarchy and enforced pregnancy.[8] Some 35-55 million abortions take place world-wide each year (Ewart and Winikoff, 1998). Abortion has become a taken-for-granted aspect of our culture. Similarly, embryo stem cell research has been heralded as very likely leading to significant new treatments for diabetes and Parkinson's disease, immunodeficiencies, cancer, metabolic and genetic disorders, and a wide variety of birth defects, as well as being useful for generating new organs or tissues.[9] As the NIH anticipates, the necessary embryos are typically generated through in vitro fertilization in the laboratory strictly for research purposes or are leftovers from in vitro fertilization fertility treatments. Since such basic science will likely save lives, reduce suffering, and help cure disease, from a general secular perspective, it appears decidedly immoral to fail to engage in such research.[10] The dominant secular culture does not regard the early embryo, much less the unborn child as a person (see *Roe v. Wade*, 1973).

This circumstance is not simply the result of an intellectual mistake; *pace* May and Colvert, the intrinsic value of embryonic life is not understandable through secular reason. Consider the conclusions of Leon Kass, chairman of the President's Council on Bioethics:

> Granting that a human life begins at fertilization and develops via a continuous process thereafter, surely, – one might say – the blastocyst itself can hardly be considered a human being. I myself would agree that a blastocyst is not, in a *full* sense, a human being – or what the current fashion calls, rather arbitrarily and without clear definition, a person. It does not look like a human being nor can it do very much of what human beings do. Yet, at the same time, I must acknowledge that the human blastocyst is (1) human in origin and (2) *potentially* a mature human being, if all goes well. This, too, is beyond dispute; indeed it is precisely because of its peculiarly human potentialities that people propose to study it rather than the embryos of mammals (Kass, 2002, p. 88).

Moreover, in their report on reproductive technologies, the President's Council announced that definitively determining the moral standing of human embryos presented intractable differences: "Like the nation at large, our members hold differing views about certain foundational questions, especially the moral standing of human embryos" (2004, p. xvii). They concluded that

> ...there is deep disagreement in our society about the degree of respect owed to in vitro embryonic human life and the weight that respect should carry in relation to other moral considerations, such as helping infertile couples to have children, helping couples to have healthy children, and advancing biomedical knowledge that could well lead to cures for dread diseases (2004, p. 8).

Whereas May and Colvert voice strong concerns regarding the protection of innocent human life and the intrinsic value of the human embryo, for Kass, many members of the President's Council on Bioethics, and much of the American public, the status of the embryo and its permissible use in medical research just appear different. Even if all agreed that intrinsically evil acts violate and radically contradict the good of the person (to paraphrase May), we disagree about which acts are, in fact, intrinsically evil. Secular reason is simply inadequate to the challenge of securing a traditional Christian bioethics and medical morality.

Articulating an answer requires an actual determination of facts as well as specification of the criteria and standards of rational choice employed.[11] Even to separate information from noise one must specify standards of evidence and inference. As a result, securing a particular account of human nature and its core goods requires a specification of moral content that will not be acknowledged as authoritative among those who do not already share a common moral anthropology. Each will make similar, equally strongly held, but contradictory claims regarding objective truth. Privileging one moral anthropology over others as the ground of secular reasoning straightforwardly begs the question. Whereas it may be true that "…we know, deep in our hearts, that we are called to seek the truth about what we are to do, to cleave to it once we have discovered it, and to shape our choices, our actions, and our lives in accord with it" (May, 2004, p. 38, quoting John Paul II), since all do not hear the same god, much less hear the one true God the same way, all do not find the same truth deep in their hearts.

3. PLACING BIOMEDICAL CHOICES WITHIN A CHRISTIAN LIFE

An essential component for judging the morality of biomedical choices, an element inadequately developed within May's and Colvert's essays, is an account of the proper role of medicine in a Christian life. Modern medicine continually offers new and costly diagnostic and therapeutic advancements which are supportive of obvious human goods. The increasingly significant investment of personal and social resources into high technology medicine is driven by the very real possibilities of ameliorating the physiological collapse brought on by age, accident, injury, and disease. Such opportunities and temptations must be placed within and appreciated in terms of a fully Christian life. Here, I draw on St. Basil the Great (A.D. 329-379), St. Maximos the Confessor (A.D. 580-662), and St. Isaac the Syrian (A.D. 613-?) as sources for reassessing the morality of medicine and the acquisition of rightly directed moral knowledge. These sources are chosen because they lie at the roots of traditional Christian morality – the historic roots of Roman Catholic natural law moral reflection – against which the contemporary culture of death defines itself. Their reflections on medicine, the human good, and its relationship to worship, spiritual therapy, and God will be used as a basis to indicate a broader philosophical perspective, which will be needed to avoid a one-sided, incomplete approach to the challenges of addressing John Paul II's diagnosis of our cultural crisis.

Writing in the fourth century, St. Basil notes that medicine is an important good to be used to relieve sickness and suffering: "Each of the arts is God's gift to us, remedying the deficiencies of nature … the medical art was given to us to relieve the sick, in some degree at least" (1962, pp. 330-331). Medical developments do not, however, change the primary undertaking of traditional Christianity: man's relationship with God. As with all human goods, medicine must be placed within the Christian life and appreciated in terms of a therapeutic reorienting of the heart towards God. In this vein, St. Maximos reminds us that it is the distortion of such goods, rather than the goods themselves, which is sinful: "It is not food that is evil but gluttony, not the begetting of children but unchastity, not material things but avarice, not esteem but self-esteem. This being so, it is only the misuse of things that is evil..." (St. Maximos, 1981, p. 83). Similarly:

> Both spiritual knowledge and health are good by nature, yet their contraries have been of more benefit to many people. For such knowledge may serve no good purpose where the wicked are concerned, even though, as we have said, it is good in itself. The same is true with regard to health, riches and joy, for they are not used advantageously by such people. But certainly their contraries do benefit them. Therefore not one of them is evil in itself, even though it may appear to be evil (St. Maximos, 1981, p. 78).[12]

Medical and therapeutic interventions cannot be adequately judged if one regards only bodily functioning, pain and suffering, or liberty, equality and justice interests, without direct and immediate reference to one's relationship to others and to God (see generally Hughes, 2002). Medicine must be appreciated within a fuller spiritual context.

Within traditional Christian reflection, for example, the evil of abortion is recognized as having a moral and spiritual impact equivalent to murder. The *Didache*, which dates from the first century A.D., states: "Do not murder a child by abortion, nor kill it at birth (Sparks, 1978a, p. 309). Likewise, the *Epistle of Barnabas*, dated to the first or second century A.D.: "Do not murder a child by abortion, nor, again, destroy that which is born (Sparks, 1978b, p. 298). Canon XXI of the regional council of Ancyra (A.D. 314) affirmed:

> Concerning women who commit fornication, and destroy that which they have conceived, or who are employed in making drugs for abortion, a former decree excluded them [from Holy Eucharist] until the hour of death, and to this some have assented. Nevertheless, being desirous to use somewhat greater lenity, we have ordained that they fulfill ten years [of penance], according to the prescribed degrees (Schaff and Wace, 1995, second series, volume XIV, p. 74).

Canon 91 of the Quinisext Council (A.D. 691) similarly decreed: "Those who give drugs for procuring abortion, and those who receive poisons to kill the fetus, are subjected to the penalty of murder" (Schaff and Wace, 1995, second series, volume XIV, p. 404). Moreover, as Canon II of Basil the Great makes clear, the ensoulment or state of formation of the fetus is irrelevant to the Church's judgment of abortion: "Let her that procures abortion undergo ten years' penance, whether the embryo were perfectly formed, or not" (St. Basil, 1995, vol. XIV, p. 604). To understand abortion rightly, it must be seen in terms of its full spiritual implications. St. Basil thus recognizes that even early embryocide has the same spiritual effect as murder, without ever committing himself to an understanding of the embryo either as a small person or as already possessing a soul. Such a moral proscription cannot be fully appreciated through secular philosophy.

In short, one must not treat materialistically what is essentially a spiritual reality. Properly appreciated and directed, medicine's role can be both physically and spiritually therapeutic. For the traditional Christian, sickness and suffering, pain and debilitation, can be positive goods, if used as a means of communion with Christ.[13] Suffering is not sought after for its own sake. Indeed, St. Basil affirms the appropriateness of administering analgesics to control pain: "...with mandrake doctors give us sleep; with opium they lull violent pain (St. Basil, 1994, p. 78). Yet, if appropriately addressed, so that it does not lead to despair, suffering reminds one to learn to control one's passions, to repent of one's transgressions, unselfishly to love others, and to turn to God. Still, it is not by suffering that one is blessed. It is only suffering for the sake of Christ and after His example that orients one towards God.[14] Within a fully Christian context, suffering reminds us to treat the soul as well as the body. "Since the soul is more noble than the body and God incomparably more noble than the world created by Him, he who values the body more than the soul and the world created by God more than the Creator Himself is simply a worshipper of Idols" (St. Maximos, 1981, p. 53). In this way, disability and illness, pregnancy and the raising of children, can help one to love others more than oneself, to learn humility and to orient oneself towards God, thereby therapeutically treating the effects of sin.

Safeguarding one's body and life are obligations to oneself, others, and the Creator. Significant duties to oneself and others, such as to one's spouse and children, generally require that one seek and accept appropriate medical therapy. Health care and nourishment are valuable means for preserving bodily health and thereby for fulfilling one's role in this life. As already noted, medicine is an important good to be used to cure illness, treat disease, and relieve suffering. Yet, if the pursuit of health and temporal life becomes an all-absorbing project, medicine distracts one from the cardinal human goal that lies beyond this world: the Kingdom of God (Engelhardt, 2000, p. 318). Here St. Basil reminds us: "Whatever requires an undue amount of thought or trouble or involves a large expenditure of effort and causes our whole life to revolve, as it were, around solicitude for the flesh..." should be avoided (1962, p. 331). Similarly, St. Maximos notes: "If you distract your intellect from its love for God and concentrate it not on God, but on some sensible object, you thereby show that you value the body more than the soul and the things made by God more than God Himself" (1981, pp. 54-55). Medicine only postpones death; it ameliorates base human suffering, but does not cure death. Medical interventions may be permissibly withheld or withdrawn insofar as they regard only materialistic solicitude for this life, disordering basic human goods and one's relationship to God, thereby leading to spiritual disorientation. Similarly, to emphasize, abortion is not condemned because one can demonstrate in secular philosophical terms that the embryo or fetus is equivalent to a human adult; rather it is condemned because of its immense spiritual significance as equivalent to murder.[15]

4. TRADITIONAL MORAL EPISTEMOLOGY

While May and Colvert strive to reorient human politics and personal morality in Roman Catholic terms through the natural law, their focus on rational philosophical argument misses the cardinal element of Christian anthropology: human beings are created to worship God. Morality appropriately understood must be supportive of and incidental to this cardinal, ontological characteristic. Moral action, true faith, and right worship are intimately interwoven. Christian moral epistemology must be nested within this cardinal element of traditional Christian anthropology. To know truly regarding medical morality and bioethical decision-making, one must correctly orient oneself towards God; that is, one must worship rightly. Incorrect worship and false belief can lead to significant spiritual disorientation even for sincere and conscientious Christians. As Romans 1:20-31 makes clear, the failure to worship correctly leads also to a darkening of the heart; a failure to know truly regarding our moral and spiritual obligations. The fundamental root of Christian bioethics is the real experience of the fully transcendent God.

As St. Isaac the Syrian observes, it is by opening our hearts to God that we come to know rightly. He likens this knowing to possessing two types of sight: the view of physical objects and the vision of spiritual perception.

> What the bodily eyes are to sensory objects, the same is faith to the eyes of the intellect that gaze at hidden treasures. Even as we have two bodily eyes, we possess two eyes of the soul, as the Fathers say; yet both have not the same operation with respect to divine vision. With one we see the hidden glory of God which is concealed in the natures of things; that is to say, we behold His might, His wisdom, and His eternal providence for us which we understand by the magnitude of His governance on our behalf. With this same eye we also behold the glory of His holy nature. When God is pleased to admit us to spiritual mysteries, He opens wide the sea of faith in our minds … (1984, Homily 46, p. 223).

Moreover,

> The intellect is spiritual perception that is conditioned to receive the faculty of divine vision, even as the pupils of the bodily eyes in which sensible light is poured. Noetic vision is natural knowledge that is used [by power] to the natural state and it is called natural light (1984, Homily 66, p. 323).

The process of traditional Christian spiritual therapy seeks noetic understanding through rightly oriented asceticism, prayer, and liturgy. This knowledge of the heart, the soul's knowledge of the inner essence of reality, is not acquired by rational philosophical argument, scientific study, or political action, but by a life strengthened through fasting, almsgiving, and prayer, in which one learns to turn fully to God.

Appeal to the deep moral insights and reflections of individual conscience has come to represent independent access to moral, theological, and spiritual truth. May and Colvert recognize that such an appeal is hazardous, unless placed within the requirements of an objective moral order. The significant danger of such an appeal, from a traditional Christian perspective, is that if one is misled by one's passions, such as despair, anger, or hatred, spiritually disordered desires, subtle terminological shifts, or an errant underlying intellectual and moral culture, one's conscience will choose evil, all the while affirming such evil as good. Even John Calvin warned his protestant followers cautiously to assess one's conscience. He noted that "... the pagans say that true glory consists in an upright conscience. Now, this is true, but it is not the whole truth. Since all men are blinded by too much self-love, we are not to be satisfied with our own judgment of our deeds. We must keep in mind what Paul says elsewhere: that even though he is not aware of anything [wrong] in him, he is not therefore justified (1 Cor. 4:4)" (Commentaries, 1958, Chapter 8). The background moral and intellectual culture affirms practices and lifestyles, such as abortion, embryocide, and embryo research, that are from a traditional perspective sinful, sustaining selfish and inappropriate passions, thereby darkening man's intellect and his capacity to know truly. It is thus often difficult to discern good from evil undistorted by desire. This is why straightforward appeal to the moral judgments of one's conscience is misguided.

Persons are at liberty to sin. However, freely chosen and conscientiously willed sin is still evil. Even if one sins in ignorance or by mistake, it remains sin. Consider by way of analogy a physician who in ignorance or through error gives a transfusion of blood tainted with HIV or other blood borne pathogen. Just as such a transfusion will harm the patient, sinning in ignorance or by mistake still causes spiritual harm. It will thus require appropriate spiritual therapy and guidance. Appeal to the moral primacy of the intuitions of one's conscience, from a traditional perspective can only be appreciated as a right to pervert one's own heart, affirming one's own will, over against God's, as "good". In this vein, one can appreciate Milton's Satan in *Paradise Lost*, who guided by passionate hatred and perverse desire proclaims: "Evil be thou my Good".[16]

Traditional Christian moral epistemology is a process of repentance and illumination. Spiritual knowledge comes as one through ascetic discipline acquires a will in union with God, rather than in union with one's own passions and desires.

> Natural knowledge, which is the discernment of good and evil implanted in our nature by God, persuades us that we must believe in God, the Author of all. Faith produces fear in us, and fear compels us to repent and to set ourselves to work. And thus man is given spiritual knowledge, which is the perception of mysteries, and this perception engenders the faith of true divine vision. Spiritual knowledge is born, just as Saint John Chrysostom has said, 'For when a man acquires a will that conforms to the fear of God and to right thinking, he quickly receives the revelation of hidden things.' And by 'revelation of hidden things' he means spiritual knowledge (1984, Homily 47, p. 227).

Moral knowledge is acquired through spiritually therapeutic practice. Engaging in action which changes oneself and leads one closer to God, treats the soul and cures the effects of sin so that one can learn to judge rightly.

Morality, the good life, and the acquisition of spiritual virtue serve to change the person to allow the acquisition of knowledge which goes beyond the individual to God. Human goods, true human flourishing, must be reordered in terms of Divine reality. To know the good, to comprehend true human flourishing, one must be oriented correctly towards God, spiritually healthy, and united with Him. There is, as Engelhardt puts it, "...a synergy of the Creator with the creature, through which God reaches out to those who reach to Him as they turn to him..." (2000, p. 183, see also chapter 4 generally). Consider Maximos the Confessor:

> The soul would never be able to reach out toward the knowledge of God if God did not allow himself to be touched by it through condescension and by raising it up to him. Indeed, the human mind as such would not have the strength to raise itself to apprehend any divine illumination did not God himself draw it up, as far as is possible for the human mind to be drawn, and illumine it with divine rightness (1985, p. 134).

What is at stake is not the adequacy of a philosophical system or appropriately identifying, defining, and ranking of central values, such as "human dignity" and "solidarity", or "social justice"; but rather, the experience of God. Christian moral epistemology must, therefore, consider whether those who are held to be exemplar knowers of God's will have turned away from God towards themselves and their own desires and passions, or if they have instead turned fully to God (Cherry, 1996; 2002).

It is a consideration of the relationship between Creator and creatures, where one cures oneself of the effects of sin through ascetic discipline, and learns to pray rightly, and thereby comes to experience the Creator. As St. Isaac observes, the experience of such an encounter with the infinite and transcendent is such that those who experience God:

> ...can soar on wings in the realms of the bodiless and touch the depths of the unfathomable sea, musing upon the wondrous and divine workings of God's governance of noetic and corporeal creatures. It searches out spiritual mysteries that are perceived by the simple and subtle intellect. Then the inner senses awaken for spiritual doing, according to the order that will be in the immortal and incorruptible life. For even from now it has received, as it were in a mystery, the noetic resurrection as a true witness of the universal renewal of all things (1984, Homily 52, p. 261).

True moral knowledge requires a change in the knower, so that the knower can learn to orient himself rightly and experience God. The goal of such a life change is not virtue or goodness as ends in themselves, but holiness. Thus traditional Christianity recognizes as exemplar knowers of God's will those who have fully repented, purified their hearts from their passions, turned fully to God, and have thereby become holy. Such persons are Saints, not merely as symbols of persons who have led exemplary or good lives, but as persons who are truly in living communion with God. The center of the moral life is, therefore, right worship, where the goals of morality and bioethics are spiritually therapeutic. Here the central epistemic vantage point is to be found in the union of the Christian assembly in the Liturgy, where liturgical prayer is the union of the community with God.

Christian morality is first and foremost a way of life rooted in worship and asceticism directed toward a personal and loving God. Only in this context can one come to know truly how one ought to choose to live. It is a turning of the heart fully to God.[17] Medicine's ability to treat somatic and psychological illnesses is only properly appreciated when that use is placed within the larger context of curing the soul. Disengaged from a way of life grounded in right worship and the development of ascetic discipline directed toward the therapeutic curing of the soul, which orient a traditional Christian life, it is difficult to appreciate what might be wrong with abortion or the research use of embryos. Yet, in such choices one subverts God's will with one's own.

5. CONCLUDING THOUGHTS: THE NEED FOR A RADICAL CURE

May and Colvert each rightly note that there is a tendency within the pluralistic democracies of the west to reject a relationship between civil authority and a substantive conception of Christian moral good. Each reflects John Paul II's hope to unify democratic politics, human dignity, human rights, and social justice, with an objective Roman Catholic moral order:

> John Paul's answer to this concern is that there is no insurmountable difficulty with democratic politics and the language of rights when both are grounded in a prior objective moral order. This order, specified by the principles of the natural law, is the necessary basis for civil society (Colvert, 2004, p. 58).

> The great truth that absolute moral norms proscribing intrinsically evil acts are "valid always and for everyone, with no exceptions," is essentially related to the truth that human persons possess an inviolable dignity. In fact, as John Paul II observes, these norms "represent the unshakable foundation and solid guarantee of a just and peaceful human coexistence, and hence of genuine democracy, which can come into being and

> develop only on the basis of the equality of all its members, who possess common rights and duties" (May, 2004, p. 42).

Donum Vitae similarly argues against the authority of civil authorities to legalize reproductive techniques that interfere in the rights of Christian spouses (see Colvert, p. 9). And, *Evangelium Vitae* calls on governments to instantiate Roman Catholic understandings of life and dignity:

> This task is the particular responsibility of civil leaders. Called to serve the people and the common good, they have a duty to make courageous choices in support of life, especially through legislative measures. … But no one can ever renounce this responsibility, especially when he or she has a legislative or decision-making mandate, which calls that person to answer to God, to his or her own conscience and to the whole of society for choices which may be contrary to the common good. …I repeat once more that a law which violates an innocent person's natural right to life is unjust and, as such, is not valid as a law. For this reason I urgently appeal once more to all political leaders not to pass laws which, by disregarding the dignity of the person, undermine the very fabric of society (John Paul II, 1995, no. 90).

This substantial hope reflects a vision of Christendom – the unity of Christian truth with social and political legal structures – rather than a truly authentic guide to personal and cultural reform. The law no longer recognizes and instantiates traditional Christian goods. Indeed, as the legal developments permitting physician assisted suicide and unfettered access to abortion, third-party assisted reproduction with embryo wastage, and so forth, pay witness, the law is frequently directly hostile to traditional Christianity. The hope for Christendom is in vain.

Moreover, seeking merely to change the law as either a political challenge or a philosophical puzzle, without first and foremost seeking fully to orient oneself toward God, leads one astray from the traditional focus of the Christian life. Right worship is central to sustaining a Christian culture; liturgical unity grounds and unifies the morality and bioethics of the Christian community. It is the careful cultivation of the relationship between man and God. The intellectual and moral reformation of medical morality and bioethics depends on this more fundamental, spiritual-therapeutic approach. St. Basil the Great, St. Maximos the Confessor, and St. Isaac the Syrian address morality neither philosophically, nor legalistically, but therapeutically. Here, morality is not an academic discussion or an exercise in consensus formation; nor is it the result of discursive rational argument, or the reflections of one's individual conscience; nor can it be fully discovered or understood outside of the practice of authentic worship. Instead, medical morality and bioethics are appreciated as aspects of a living phenomenological world. Traditional Christianity understands the pursuit of a fully flourishing human life as situated within the thick expectations of family and community, liturgy and prayer, ascetic struggle and fasting. Morality is understood through this way of life rooted in right worship and right belief. Such a life orients one toward the personal and loving God Who is and medicinally treats the effects of sin.

It is for this reason that Sts. Basil and Maximos locate the role of medicine within the spiritually therapeutic turning of one's heart fully to God. Orthodox spirituality is in this sense "medicinal". It is referred to as a hospital for the curing of the soul.[18] It should be noted that Orthodox Christianity does not rely on the writings of the Saints and Fathers of the Church as independent sources of truth; but rather, Orthodoxy is immersed in the very same life-world which sustains these

Fathers and Saints (Engelhardt, 2000, p. 160). It is embedded in the very same culture sustained through unchanging practices of worship and belief. The robust, content-full vision of human flourishing within traditional Christianity cannot be captured or recreated outside of its actual practice. Entering into this life, living, believing, and worshiping as one with the ancient Christian Fathers, is essential for overcoming one-sided and inadequate approaches to morality and bioethics. Christianity is not a philosophical system,[19] a system of rights, duties, and values, or a scriptural interpretation,[20] but a living reality. The focus of theology par excellence is union with God: this is necessary for an adequate cure.

St. Basil and St. Maximos remind us that medicine can play an important role within a Christian life, but that it must not dominate one's life or become an all-consuming endeavor. St. Isaac orients the moral epistemology for Christian morality and bioethics through right worship and noetic experience of God. Medical therapy to relieve disease, distress and suffering should not replace Christian spiritual therapy for the soul. Willful termination of pregnancy, embryocide, or the destruction of human embryos for research is rejected. Such actions act directly against basic human goods and deny God's dominion over the world. While medicine is an important human good, physicians must not be understood as masters of life and death. This would be an idolatry of medicine. Traditional Christianity – that is, Christianity at one with and unbroken from, the worship and belief of the Apostles and Fathers of the Church – opposes abortion and embryocide as fundamentally incompatible with the cardinal elements of true human flourishing.

St. Edward's University
Austin, Texas

NOTES

1 For detailed descriptions of the various reproductive technologies, including fertility drugs, intrauterine insemination, egg donation, gamete intrafallopian transfer, in vitro fertilization, zygote intrafallopian transfer, surrogate motherhood, genetic surrogacy, gestational surrogacy, prenatal genetic testing, intracytoplasmic sperm injection, cloning, somatic cell nuclear transfer, and fetal egg donation see Rae (2003); for a traditional Christian analysis of many such techniques see Engelhardt (2000, especially chapter 5).

A selection of the relevant U.S. case law includes: (1) *Meyer v. Nebraska* (1923), in which the court affirmed that individual freedoms to "marry, establish a home and bring up children" were Constitutionally protected liberties (*Meyer v. Nebraska* 262 U.S. 390 [1923] at 399); (2) *Pierce v. Society of Sisters* (1925), which affirmed the liberty of parents to raise their children in the manner they think fit: "It is an unreasonable interference with the liberty of parents and guardians to direct to upbringing of the children, and in that respect violates the Fourteenth Amendment" (*Pierce v. Society o f Sisters* 268 U.S. 510 [1925], at 534); (3) *Skinner v. Oklahoma* (1942), which struck down a law that required mandatory sterilization for habitual criminals. The Court opined: "We are dealing here with legislation which involves one of the basic civil rights of man. Marriage and procreation are fundamental to the very existence and survival of the race. [When sterilized] he is forever deprived of a basic liberty (*Skinner v. Oklahoma* 316 U.S. 535 [1942]); (4) *Griswold v. Connecticut* 381 U.S. 479 (1965), the Supreme Court decision that struck down a Connecticut law forbidding the use of contraceptives, and is seen as recognizing a right of privacy in marriage; (5) *Eisenstadt v. Baird* 405 U.S. 438 (1972), which broadened the decision of *Griswold* to include unmarried individuals as well as married individuals; (6) *Carey v. Population Services International* (1977), which affirmed a lower court decision to strike down a New York law that restricted the sale of contraceptives to minors, required the purchase of contraceptives from a licensed pharmacist, and prohibited the advertising of contraceptives. The court ruled: "The decision to bear or begat a child is at the very heart of this cluster of constitutionally protected choices. That decision holds a particularly important place in the history of the right to privacy. Decisions whether to accomplish or prevent conception are among the most private and sensitive" (*Cary v. Population Services International* 431 U.S. 678 [1977], at 685). For additional discussion of such case law see Rae, 2003.

2 For a set of religious reflections on the moral permissibility of abortion, recast rhetorically as "sacred choices" see Maguire, 2001.

3 As Boyle makes the point: "Health is good bodily functioning and is the perfection of our being alive just as animals. Being alive is not only a necessary condition for pursuing other goods but is part of the personal reality of a human being. This means that life and health are intrinsically good; that is, whenever an action in view promotes or protects life or health, that action is so far forth choice-worthy; that it protects or enhances life or health provides a reason sufficient for doing it, though not necessarily a reason that morally justifies doing it" (Boyle, 2002, p. 79).

4 Germain Grisez's explication is helpful: "Many intelligible goods are only instrumental: One diets to reduce cholesterol, looks both ways to avoid getting hit, brushes one's own teeth to remove plaque. Lowering cholesterol, not getting hit, and removing plaque are indeed reasons for acting, but are intelligibly good only because they contribute to staying alive, intact, and healthy. Thus, free choices to diet, look both ways, and brush presuppose insight into the truth that life, including bodily integrity and health, is a good to be protected and promoted. That truth is a principle of practical reason, and the human good to which it directs action is basic. In other words, the prospect of benefits in respect to survival, bodily integrity, or health can be one's ultimate reason for choosing. Of course, one also can regard basic goods as means to other goods" (Grisez, 2001, pp. 4-5).

5 According to Finnis, Boyle, and Grisez, the basic human goods include : "As *animate*, human persons are living organic substances. Life itself – its maintenance and transmission – health, and safety are one form of basic human good. …

As *rational*, human beings can know reality and appreciate beauty and whatever intensely engages their capacities to know and feel, and to integrate the two. Knowledge and aesthetic experience are another category of basic good.

As simultaneously *rational* and *animal*, human persons can transform the natural world by using realities, beginning with their own bodily selves, to express meanings and/or serve purposes within human cultures. Such bestowing of meaning and value can be realized in diverse degrees; its fullness is another category of basic good: excellence in work and play" (1987, p. 279).

The basic human goods transcend any particular state of affairs, are never fully attainable, require no further or prior reasons (i.e., they are intrinsically valuable), provide reasons for other objectives, are non-contingent, components for full human flourishing, and non-competitive, i.e., one person's pursuing the good does not thereby preclude others from pursuing the same good" (Finnis, Boyle, and Grisez, 1987, pp. 277-278).

6 As Ana Iltis has argued, even within a Christian context, there are at least four different human natures which compete to ground natural law: "Humans may generally be said to have four natures: (1) human nature before the Fall; (2) human nature after the Fall; (3) redeemed human nature; and (4) restored human nature. Thus even if we grant the assumption natural law morality makes, namely that humans share one normative nature whose ends are knowable and morally good, we must recognize that there are four natures that might be normative" (Iltis, 2004, p. 115).

7 Consider the extensive, and frequently non-Christian, positive human rights enumerated in various statements of the United Nations. For example, the United Nations published guidelines on "HIV/AIDS and Human Rights" (1996), which specifically calls for unfettered access to abortion: "Laws should also be enacted to ensure women's reproductive and sexual rights, including the right of ... means of contraception, including safe and legal abortion and the freedom to choose among these, the right to determine number and spacing of children..." (pp. 20-21). Failing to provide safe and legal abortion as a matter of positive right is understood as a violation of basic human rights. Similarly, the United Nations' International Bioethics Committee concluded favorably regarding pre-implantation genetic diagnosis, with embryo wastage to avoid giving birth to children with genetic abnormalities (Galjaard, 2003). The draft report of the International Bioethics Committee on the Possibility of Elaborating a Universal Instrument on Bioethics openly acknowledges the likely moral permissibility of utilizing human embryos in research (International Bioethics Committee, 2003).

8 As Suzanne Poppema encapsulates: "I'm an abortion doctor…What I do is right and good and important. Perhaps my story will appall some, but it also may inspire others, particularly the young

women who need to know that the struggle between feminism and the patriarchy has not been in vain" (1996, p. 11). Similarly, Todd Whitemore, a moral theologian at Notre Dame University, states: "I understand oppression to mean any social situation that systematically requires heroism on the part of a class of people based on gender, race or other such identification. Given the gendered structure of society at present and the resultant lack of communal support for women, the absolute prohibition on taking the life of the fetus, however protective of the right to life of that fetus, remains oppressive of women" (1994, p. 14).

9 The United Kingdom has granted embryo research licenses. As Mayor documents, human embryonic stem cells are useful because they are very primitive – they have the potential to develop into any type of cell in the body. Stem cell therapy is believed to be effective in treating diseases such as diabetes and Parkinson's, but it has been difficult to produce enough cells to treat even one patient (Mayor, 2002).

The American Academy of Pediatrics Committee on Pediatric Research and Committee on Bioethics has also recently endorsed human embryo research: "Recently, several investigators have successfully isolated and cultured pluripotent stem cells from frozen human embryos donated by couples who had previously undergone in vitro fertilization and whose additional embryos were no longer clinically needed. Concrete benefits for children resulting from pluripotent stem cell research with human embryos are anticipated, including treatments for spinal cord and bone injuries, diabetes, primary or acquired immunodeficiencies, cancer, metabolic and genetic disorders, and a variety of birth defects. Research with human embryos that involves drug or toxin testing could benefit children suffering from toxicities of drug treatment and environmental pollutants as well as prenatal drug use disorders, such as fetal alcohol syndrome. Research using material derived from embryos also could be used in the study of normal and abnormal differentiation and development, which could benefit children with birth defects, genetically derived malignancies, and certain genetic disorders.

An important long-term benefit of research using human embryos can be found in the field of teratology. Experiments that involve exposing pregnant mice to a teratogen at specific times after mating and observing the resulting defects have demonstrated that early exposure can result in very specific developmental defects. Other research on mouse embryos has advanced the ability to evaluate gene function in the early embryo. In the future, it may be possible to combine these approaches to obtain important insights into teratogenesis in humans. Specifically, there is the future prospect of studying the expression of specific genes in embryonic cell lines exposed in vitro to teratogens. Such research could potentially provide insights into the approximately 40% of anatomic defects in infants for which there are currently no explanations" (2001, p. 814).

10 The National Institute of Health's Human Embryo Research Panel issued a report concluding that research on pre-implantation human embryos was morally permissible, as was the creation of embryos for such research utilizing *in vitro* fertilization techniques (NIH, 1998 [1994]). The NIH anticipated that gametes and embryos could be obtained from the following sources: Women/couples in IVF programs, women undergoing scheduled pelvic surgery, women undergoing pelvic surgery for research that involves transfer of the resulting embryo for the purpose of establishing a pregnancy, women who are deceased provided that the particular woman had not expressly objected to such use of her oocytes and that the next of kin has consented (NIH, 1998 [1994]). The panel argued that the human embryo "warranted serious moral consideration as a developing form of human life" (p. 932), but that it does not have the same moral status as a person. On the one hand, such research has been denounced as gravely immoral by many Christians: "After carefully studying the Report on the Human Embryo Research Panel, we conclude that this recommendation is morally repugnant, entails grave injustice to innocent human beings, and constitutes an assault upon the foundational ideas of human dignity and rights essential to a fee and decent society" (The Ramsey Colloquium, 1995, p. 17). Similarly: "To use human embryos or fetuses as the object or instrument of experimentation constitutes a crime against their dignity as human beings having a right to the same respect that is due to the child already born and to every human person" (*Donum Vitae*, 1987, no. I, 4).

Donum Vitae spoke to IVF and cloning as well: "Techniques of fertilization in vitro can open the way to other forms of biological and genetic manipulation of human embryos, such as attempts or plans for fertilization between human and animal gametes and the gestation of human embryos in the uterus of animals, or the hypothesis or project of constructing artificial uteruses for the human embryos. These procedures are contrary to the human dignity proper to the embryo, and at the same time they are contrary to the right of every person to be conceived and to be born within marriage and from marriage. Also, attempts or hypotheses for obtaining a human being without any connection with sexuality through 'twin fission,' cloning or parthenogenesis are to be considered contrary to the moral law, since they are in opposition to the dignity both of human procreation and of the conjugal union" (*Donum Vitae*, 1987, no. I, 6). Similarly, "What is morally unsustainable is the harvesting of stem cells by either of two currently proposed methods: 1) the creation and destruction of human embryos at the blastocyst stage by removal of the inner cell mass or 2) the harvesting of primordial germ cells from aborted fetuses. Both cases involve complicity in the direct interruption of a human life, which Roman Catholics believe has a moral claim to protection from the first moments of conception. In both cases, a living member of the human species is intentionally terminated" (Pellegrino, 2000, p. F3).

In contrast, Orthodox Jewish scholars have expressed the view that human beings are commanded to use their rational capacities to overcome and master natural limitations. In the context of embryo stem cell research, for example, Rabbi Moshe Tendler argued: "The commendable effort of the Catholic citizens of our country to influence legislation that will assist in preventing the further fraying of the moral fabric of our society must not impinge on the religious rights and obligations of others. Separation of church and state is the safeguard of minority rights in our magnificent democracy. Life-saving abortion is a categorical imperative in Jewish biblical law. Mastery of nature for the benefit of those suffering from vital organ failure is an obligation. Human embryonic stem cell research holds that promise. The recently announced joint effort between Geron Corporation and Roslyn Labs of Scotland (of Dolly fame) has its focus on the use of human embryonic stem cells to bolster a failing heart or liver, without need for immunosuppressive drugs or dependency on organ donors" (Tendler, 2000, p. H4). Jewish scholars have thus been wary of public policy prohibiting all human embryo research because it would violate the command of mastery, contravene the separation of church and state, and interfere with potentially valuable scientific research. As Courtney Campbell documents this Jewish position: "Human beings have a command and challenge from God to use their rational, imaginative, and exploratory capacities for the benefit and health of humanity. Judaism affirms that human beings have inherent worth as creatures created in the image of God, and the Talmud understands human beings as partners with God in the ongoing act of creation. In their unique role, persons receive a divine mandate for stewardship and mastery, which encompasses a very strong emphasis on use of medical knowledge and skills to promote health, cure, and heal" (Campbell, 1997, p. D29).

11 Here one might consider Alasdair MacIntrye's reflections on the insurmountable challenges of articulating a universal morally authoritative account of practical rationality. "Yet, someone who tries to learn this at once encounters the fact that disputes about the nature of rationality in general and about practical rationality in particular are apparently as manifold and as intractable as disputes about justice. To be practically rational, so one contending party holds, is to act on the basis of calculations of the costs and benefits to oneself of each possible alternative course of action and its consequences. To be practically rational, affirms a rival party, is to act under those constraints which any rational person, capable of an impartiality which accords no particular privileges to one's own interests, would agree should be imposed. To be practically rational, so a third party contents, is to act in such a way as to achieve the ultimate and true good of human beings" (1988, p. 2). Should one seek to balance costs and benefits so as to maximize the good with John Stuart Mill (1988 [1861]) or Peter Singer (1993)? If so, whose account of costs and benefits should be utilized: traditional Moslems, Orthodox Christians, or secular atheists? Should one accept the judgments of fully rational contractors under a veil of ignorance with John Rawls (1999), or the rational libertarianisms of Robert Nozick (1974) and David Friedman (2000)? One finds a plurality of standards of practical reason.

12 Similarly: "Just as the intellect of a hungry man imagines bread and that of a thirsty man water, so the intellect of a glutton imagines a profusion of foods, that of a sensualist the forms of women, that of a vain man, worldly honour, that of an avaricious man financial gain, that of a rancorous man revenge on whoever has offended him, that of an envious man how to harm the object of his envy, and so on with all the other passions" (St. Maximos, 1981, p. 76).

13 As Father Edward Hughes comments: "The mortality of the body: sickness, increasing weakness, and eventual death all can be an ultimate terror, or perhaps something else. In light of the Christian tradition, the mortality of the body holds terror only for those whose lives were lived outside of Christ, or in rejection of Him and his Way. For the faithful believer, the mortality of the body leads only to a temporary separation of soul and body, in the loving care of God, and not a separation from Him or from the Christian community" (Hughes, 2002, p. 248).

14 "There are many people in the world who are poor in spirit, but not in the way that they should be; there are many who mourn ... many are gentle, but towards unclean passions; many hunger and thirst, but only to seize what does not belong to them and to profit from injustice; many are merciful, but towards their bodies and the things that serve the body; many are pure in heart, but for

the sake of self-esteem; many are peace-makers, but by making the soul submit to the flesh; many are persecuted, but as wrongdoers; many are reviled, but for shameful sins. Only those are blessed who do or suffer these things for the sake of Christ and after His example" (St. Maximos, 1981, p. 90).

15 As this traditional approach to abortion is spiritually oriented, it sought even to avoid involuntary homicide or proximal involvement in homicide. As a result, it did not draw a straightforward line between abortion and miscarriage. As Engelhardt notes: "There has been an appreciation that even an involuntary involvement in homicide can harm the human heart. As a consequence, the absolution for miscarriage recognizes (1) the importance of repentance for the ways in which we are implicated in the broken and sinful character of the world, which can even involve involuntary complicity in the death of an innocent person, (2) the need to forgive any involvement on the part of the woman in the loss of her unborn child's life, and (3) the inclusion in prayerful consideration of those around the woman who may have in some way precipitated the miscarriage" (2000, p. 277). It is a recognition of the sin of involuntary homicide and of the spiritual harm which such a sin can cause.

16 "All hope excluded thus, behold in stead
Of us out-cast, exil'd, his new delight,
Mankind created, and for him this World.
So farwel Hope, and with Hope farwel Fear,
Farwel Remorse: all Good to me is lost;
Evil be thou my Good; by thee at least
Divided Empire with Heav'ns King I hold
By thee, and more then half perhaps will reigne;
As Man ere long, and this new World shall know" (Paradise Lost, Book 4).

17 "Know well that the beginning of the clear way of God, and the coming of all good things, is for a person to realize his weakness. For him to realize it though, he must undergo great temptations above his strength. Without undergoing such extraordinary temptations above nature, it is impossible for him to realize the weakness of his nature. Once he realizes it, he knows everything and everything will fall into his hands. Then true humility is near. Within it is also patience. He has also laid hold of the knowledge of mysteries and is sheltered by discernment. From Love, he has received the fruits of the Spirit: joy, peace, long-suffering, faith, meekness, abstinence" (Elder Joseph, 1998, p. 390).

18 "The Orthodox Church, however, does not just stress the necessity of cure; it also outlines the means by which it can be achieved. ... Orthodoxy is not like philosophy. It is more closely related to the applied sciences, mainly to Medicine. This is said in the sense that cure in the Orthodox Church can be substantiated by its results. These three stages of the spiritual life are in reality participation in the purifying, illuminating and deifying energy of God, for they are attained through divine grace. ... The cure of the whole person, which is the essential aim of Orthodox spirituality if effected by the sacraments of the Church and by the practice of the ascetic life" (Vlachos, 1992, p. 98).

19 As Archimandrite Vasileios comments: "Theology does not have a philosophy of its own, nor spirituality a mentality of its own, nor church administration a system of its own, nor hagiography its own artistic school. All these emerge from the same font of liturgical experience. They all function together in a Trinitarian way, singing the thrice-holy hymn in their own languages. ... Nothing in the Church is arbitrary, or isolated, or alien, or mechanically added. Nothing has a law of its own, its own "will" in the sense of rebellion. Nothing enters into it that is alien in nature, understanding or attitude. Everything is illumined by the grace of the Trinity. Each part lives with the rest in an organic unity and is embodied in the whole" (1984, p. 11).

20 St. John Chrysostom even expresses concern for the need to rely at all on the Bible: "It were indeed meet for us not at all to require the aid of the written Word, but to exhibit a life so pure, that the grace of the Spirit should be instead of books to our souls, and that as these are inscribed with ink, even so should our hearts be with the Spirit. But, since we have utterly put away from us this grace, come, let us at any rate embrace the second best course. For that the former was better, God hath made manifest, both by His words, and by His doings. Since unto Noah, and unto Abraham, and unto his offspring, and unto Job, and unto Moses too, He discoursed not by writings, but Himself by Himself, finding their minds pure" (Homily on the Gospel of Matthew, I.1, 1994, vol. 10, p. 1).

REFERENCES

American Academy of Pediatrics Committee on Pediatric Research and Committee on Bioethics (2001). 'Human embryo research,' *Pediatrics,* 108 (3), 813-817.

Aquinas, Thomas (1948). *Summa Theologiae.* Fathers of the English Dominican Province (Trans.), Westminster: Christian Classics.

Basil, Saint (1962). 'Ascetical Works: The Long Rules,' in R. J. Deferrari et al. (Eds.), *The Fathers of the Church.* Sister M.M. Wagner, C.S.C. (Trans.), Washington, D.C.: The Catholic University of America Press.

Basil, Saint (1994). 'The hexaemeron,' in P. Schaff and H. Wace (Eds.), *Nicene and Post-Nicene Fathers.* Second Series, Vol. VIII, Peabody, MA: Hendrickson Publishers. 52-107.

Basil, Saint (1995). 'The first Canonical Epistle of our Holy Father Basil,' in P. Schaff and H. Wace (Eds.), *Nicene and Post-Nicene Fathers.* Second Series, Vol. XIV, Peabody: MA: Hendrickson Publishers. 604-608.

Boyle, J. (1999). 'Personal Responsibility and Freedom in Health Matters: A Contemporary Natural Law Perspective,' in M.J. Cherry (Ed.), *Persons and Their Bodies: Rights, Responsibilities, Relationships.* Dordrecht: Kluwer Academic Publishers. 111-141.

Boyle, J. (2002). 'Limiting Access to Health Care: A Traditional Roman Catholic Analysis,' in H.T. Engelhardt, Jr. and M.J. Cherry (Eds.), *Allocating Scarce Medical Resources: Roman Catholic Perspectives.* Washington, D.C.: Georgetown University Press. 77-95.

Calvin, J. (1958). *Commentaries.* J. Haroutunian (Ed. and Trans.), Philadelphia: Westminster Press.

Campbell, C. (1997). 'Cloning Human Beings: Religious Perspectives on Human Cloning.,' National Bioethics Advisory Committee, Rockville, M.D.

Cherry, M.J.: (1996). 'Suffering Strangers: An Historical, Metaphysical, and Epistemological Non-ecumenical Interchange.' *Christian Bioethics,* 2, 253-266.

Cherry, M.J.: (2002). 'Facing the Challenges of High-technology Medicine: Taking the Tradition Seriously,' in H.T. Engelhardt Jr. and M.J. Cherry (Eds.), *Allocating Scarce Medical Resources: Roman Catholic Perspectives.* Washington, D.C.: Georgetown University Press. 19-33.

Colvert, G.T.: (2004). 'Liberty and Responsibility: John Paul II, Ethics and the Law,' in C. Tollefsen (Ed.), *John Paul II's Contribution to Catholic Bioethics.* Dordrecht: Kluwer Academic Publishers. 51-72.

Council of Ancyra (1995). 'The Canons of the Councils of Ancyra, Gangra Neocaesarea, Antioch and Laodicea – Accepted and Received by the Ecumenical Synods,' in P. Schaff and H. Wace (Eds.), *Nicene and Post-Nicene Fathers.* Second Series, Vol. XIV. Peabody, MA: Hendrickson Publishers. 60-71.

Engelhardt, H.T. (1996). *The Foundations of Bioethics, second edition.* New York: Oxford University Press.

Engelhardt, H.T. (2000). *The Foundations of Christian Bioethics.* Lisse: Swets and Zeitlinger.

Ewart, W.R. and B. Winikoff (1998). 'Toward Safe and Effective Medical Abortion.' *Science,* 281 (July 24). 520-1.

Finnis, J., Boyle, J., and G. Grisez (1987). *Nuclear Deterrence, Morality, and Realism.* Oxford: Oxford University Press.

Friedman, D. (2000). *Law's Order: What Economics has to do with Law and Why it Matters.* Princeton Princeton University Press.

Galjaard, H. (2003). 'Report of the IBC on Pre-implantation Genetic Diagnosis and Germ-line Intervention.' *International Bioethics Committee,* UNESCO, Paris.

Grisez, G. (2001). 'Natural Law, God, Religion, and Human Fulfillment.' *The American Journal of Jurisprudence,* 46, 3-36.

Hughes, E. (2002). 'The Current Medical Crisis of Resources: Some Orthodox Christian Reflections.' in H.T. Engelhardt Jr. and M.J. Cherry (Eds.), *Allocating Scarce Medical Resources: Roman Catholic Perspectives.* Washington, D.C.: Georgetown University Press. 237-262.

Iltis, A. (2004). 'An Assessment of the Requirements of the Study of Natural Law.' in M.J. Cherry (Ed.), *Natural Law and the Possibility of a Global Ethics.* Dordrecht: Kluwer Academic Publishers. 115-122.

International Bioethics Committee, Working Group on the Possibility of Elaborating a Universal Instrument on Bioethics (2003). Draft Report. Paris: United Nations, Educational, Scientific, and Cultural Organization.

Isaac the Syrian, Saint (1984). *The Ascetical Homilies of Saint Isaac the Syrian.* Holy Transfiguration Monastery, (Trans.), Boston: Holy Transfiguration Monastery.
John Chrysostom, Saint (1994). 'Homily on the Gospel of Matthew, I.' in P. Schaff and H. Wace (Eds.), *Nicene and Post-Nicene Fathers.* Second Series, Vol. XIV. Peabody, MA: Hendrickson Publishers. 1-9.
John Paul II (1993). *Veritatis Splendor.* Vatican City: Libreria Editrice Vaticana.
John Paul II (1995). *Evangelium Vitae.* Vatican City: Libreria Editrice Vaticana.
Joseph, Elder the Hesychast (1998). *Monastic Wisdom: The Letters of Elder Joseph the Hesychast.* Florence, AZ: Saint Anthony's Greek Orthodox Monastery.
Kass, L. (2002). *Life, Liberty, and the Defense of Dignity.* San Francisco: Encounter Books.
Lustig, B.A. (2004). 'Natural Law and Global Bioethics.' in M.J. Cherry (Ed.), *Natural Law and the Possibility of a Global Bioethics.* Dordrecht: Kluwer Academic Publishers. 124-140.
Maguire, D.C. (2001). *Sacred Choices: The Right to Contraception and Abortion in Ten World Religions.* Minneapolis: Fortress Press.
MacIntyre, A. (1988). *Whose Justice? Which Rationality?* Notre Dame: University of Notre Dame Press.
Maximos, Saint (1981). 'Four Hundred Texts on Love.' in St. Nikodimos and St. Makarios (Compilers), *The Philokalia,* G.E.H. Palmer, P. Sherrard, and K. Ware (Trans.), London: Faber and Faber.
Maximos, Saint (1985). 'Chapters on Knowledge.' in G.C. Berthold (Trans.), *Selected Writings.* New York: Paulist Press.
May, W.E. (2004). 'Pope John Paul II's Encyclical *Veritatis Splendor* and Bioethics.' in C. Tollefsen (Ed.), *John Paul II's Contribution to Catholic Bioethics.* Dordrecht: Kluwer Academic Publishers. 35-50.
Mayor, S. (2002). 'United Kingdom Grants First Embryo Research Licenses.' *British Medical Journal.* 324 (7337), 562.
Mill, J,S, (1988). *Utilitarianism.* George Sher (Ed.), Indianapolis: Hackett Publishing Company.
Milton, J. (1667). *Paradise Lost.* [On-line.] Available: http: //www. literature.org /authors /milton-john/paradise-lost/index.html
National Institute for the Humanities (1994). 'Final Report of the NIH Human Embryo Research Panel.' reprinted in S. Lammers and A. Verhey (Eds.), *On Moral Medicine.* Grand Rapids: Eerdmans. 932-936.
Office of the United Nations High Commissioner for Human Rights and the Joint United National Programme on HIV/AIDS: 1996, 'HIV/AIDS and Human Rights: International Guidelines,' United Nations, Geneva, September 23-25.
Pellegrino, E. (2000). 'Testimony Provided to the National Bioethics Advisory Commission.' *Stem Cell Research, Vol. III, Religious Perspectives.* Rockville, MD: National Bioethics Advisory Commission.
Poppema, S.T. (1996). *Why I am an Abortion Doctor.* Amherst: Prometheus Books.
President's Council on Bioethics (2004). *Reproduction and Responsibility: The Regulation of New Biotechnologies.* Washington, D.C.: U.S. Government Printing Office. [On-line] Available: www.bioethics.gov.
Quinisext Council. (1995). 'The Canons of the Council in Trullo (Often called the Quinisext Council).' in P. Schaff and H. Wace (Eds.), *Nicene and Post-Nicene Fathers.* Second Series, Vol. XIV. Peabody, MA: Hendrickson Publishers. 356-409.
Rae, S.B. (2003) 'United States Perspectives on Assisted Reproductive Technologies.' in J. Peppin and M.J. Cherry (Eds.), *Regional Perspectives in Bioethics.* Lisse: Swets and Zeitlinger. 21-38.
Ramsey Colloquium, The (1995). 'The Inhuman Use of Human Beings: A Statement on Embryo Research by the Ramsey Colloquium.' *First Things,* 49, 17-21.
Rawls, J. (1999). *A Theory of Justice.* Rev. Ed. Cambridge: Harvard University Press.
Roe v. Wade (1973). 410 U.S. 113, 35 L. Ed. 2d 147, 93 S. Ct. 705.
Sacred Congregation for the Doctrine of the Faith (1987). *Donum Vitae.* Rome: Congregation for the Doctrine of the Faith.
Singer, P. (1993). *Practical Ethics.* Cambridge: Cambridge University Press.
Sparks, Jack N.: 1978a, 'The Didache,' in *The Apostolic Fathers,* Robert Kraft (trans.) Light and Life Publishing Company, Minneapolis, pp. 305-320.
Sparks, Jack N.: 1978b, 'The Epistle of Barnabas,' in *The Apostolic Fathers,* Robert Kraft (trans.) Light and Life Publishing Company, Minneapolis, pp. 266-301.

Tendler, Rabbi Moshe (2000). 'Testimony Provided to the National Bioethics Advisory Commission.' *Stem Cell Research Vol. III, Religious Perspectives*. Rockville, MD: National Bioethics Advisory Commission.

Vasileios, Archimandrite (1984). *Hymn of Entry: Liturgy and Life in the Orthodox Church.* Crestwood, NY: SVS Press.

Vlachos, Hierotheos (1992). *Orthodox Spirituality*. Levadia, Greece: Birth of the Theotokos Monastery.

Whitmore, T. (1994). 'Notes for a "New, Fresh Compelling" Statement.' *America,* 171 (10): 14.

CHAPTER SIX

LAURA L. GARCIA

PROTECTING PERSONS

1. INTRODUCTION

One of the most original contributions of Pope John Paul II to philosophy and theology is his emphasis on the moral ultimacy of persons. This moral significance is grounded in ontological truths about the nature of human beings. These truths find philosophical support in the Pope's writings, though he carefully develops their scriptural and theological foundations as well. The teaching of the Catholic Church on marriage and sexuality profits enormously from this reformulation within John Paul II's unique blend of personalism, pastoral expertise, and the perennial philosophy of Aristotle and St. Thomas Aquinas.

The personalist perspective on sex and its meaning provides justification for substantive moral claims regarding marriage and its counterfeits – co-habitation, 'serial monogamy' or multiple marriages, same-sex coupling, etc. Karol Wojtyla (John Paul II) unfolded this vision very early in his career in a series of lectures at the Catholic University of Lublin in 1958-59, ten years before Pope Paul VI issued the encyclical *Humanae Vitae* (On Human Life). In a 1960 preface to the book based on these lectures, he writes that one of its primary goals is to find a way to "introduce love into love": "The word as first used in that phrase signifies the love which is the subject of the greatest commandment, while in its second use it means all that takes shape between a man and a woman on the basis of the sexual urge" (Wojtyla, 1981, p. 17). The task is to show that the second form of love can be an incarnation of the first kind, of the love that never fails, without either downplaying the biological and psychological facts about the sexes and their mutual attraction or enfeebling our conception of genuine love.

The bedrock of this conception of love is the correct understanding of human beings as what they truly are–persons. It adds to this the fundamental insight of Immanuel Kant and others that the only adequate response to persons is love. In his philosophical works, John Paul II develops a case for this personalist norm based on human experiences of freedom in their actions and of the freedom of other human beings through their resemblance to ourselves and their purposive activity. By definition, beings that propose their own ends and purposes are persons. To *use* persons as means to an end, even a good end, is to oppose their very essence, to treat them as non-persons.[1] For John Paul II, love is the opposite of use. It acts towards other persons with a sense of their intrinsic worth, as ends in themselves, rather than as subordinate to one's own needs or desires.

C. Tollefsen (Ed.), John Paul II's Contribution to Catholic Bioethics, pp. 93–105.

In sexual ethics, the contrast between love and use can be explicated by reference to various forms of attraction between a man and a woman. On a purely physiological level, humans experience sexual urges that make a person of the opposite sex attractive or desirable. In itself, this desire is egoistic, directed at the fulfillment of the self rather than at the good of the other. At a deeper level, one might experience a kind of emotional or psychological attractiveness in a person, sometimes combined with sexual attraction to him or her. Since this emotional/passion-based attraction responds to a wider range of values in the other person than physical beauty or attractiveness, and since it still remains largely outside one's control, it can seem that this is the real essence of love, at least of erotic love. It corresponds nicely to the popular notion of 'falling in love' and to a common view on why some relationships disintegrate–the relevant feelings have disappeared.

Genuine love, as John Paul II sees it, must go beyond these attitudes, which focus mainly on the experiences of the subject and on the desirability of the other person largely as a means to these experiences. The human will can enter in here, however, to commit one to the good of another as such, desiring his or her genuine happiness and fulfillment regardless of whether this furthers one's own emotional or sexual fulfillment. If the sexual urge is to be realized in a way that enriches human persons and does not in some way abuse them, it must be taken up into this deeper commitment of the will, into betrothed love. The love envisaged here is a matter of the will more than the feelings; it requires a mutual unconditional gift of the man and woman to each other. A so-called love that dissolves when the other party ceases to bring sexual or emotional fulfillment for oneself is not true love. Hence what begins in sexual attraction and emotional feelings for another needs to develop into something much deeper if it is to be worthy of persons. This unshakeable love should be the goal of each married couple, even if it is not yet a reality in their relationship. All relationships between persons, including the marriage relationship, should be forms of true friendship–love for the other for his or her own sake.

Further insights into human nature emerge from the Pope's reflections on the creation narratives in the Book of Genesis, now collected along with related writings in a volume entitled *The Theology of the Body*. In this account of the creation of woman and of the man's first response to her ("Here, at last!" (Gn 2:23)), John Paul II finds a universal lesson and a sign of what women and men are meant to be for one another – a joy and delight, a companion, 'one like myself.' He meditates at length on the Biblical statement that, before the Fall, Adam and Eve "were naked and were not ashamed" (Gn 2:25).[2] In this state of wholly selfless love, the glory of the human body could shine forth in all its richness without becoming an occasion for possessiveness on the part of one and (consequently) fear on the part of the other. The first couple could enter into the marital act and experience all of the values of the other person as a unity–like a prism through which the light of love shines to reveal an entire range of beautiful colors. A human body is the body of a *person* and its value is transformed by that fact. Everything that affects the body affects the person and human acts take place through the body. A human body, then, is not to be treated as an object of use, just as a human person is not to be treated this way. There are not two realities here, a person and a body, but one reality–a personal, corporeal being. A human body cannot be owned or possessed any more than a person herself can be owned or possessed. The notion that human beings own their bodies is fundamentally confused. Thus, self-destructive actions are contrary to the moral imperative to respond to persons (including one's own person) with love.

2. MARRIED LOVE

The implications of this vision of love and sexuality are far-reaching indeed. We focus here on certain corollaries of the personalism of John Paul II with regard to marriage and human reproduction. Consider first the principle that contraception and sterilization are morally illicit, even within marriage. This claim strikes some today as implausible or even mildly exotic, but it was common currency among Christian thinkers until the 1920's when some Protestant denominations broke ranks and approved artificial contraceptives for married couples in certain circumstances. Catholic theologian Charles Curran of Southern Methodist University, writing on the topic of birth control for *The HarperCollins Encyclopedia of Catholicism*, notes that "before 1963 no Catholic theologian had ever publicly questioned the magisterium's teaching on artificial contraception." He observes that the present situation is quite different, and explains that "dissenting theologians [including Curran himself] disagree with the neo-Scholastic natural law theory employed by the encyclical [Pope Paul VI's 1968 encyclical *Humanae Vitae*], especially its physicalism whereby the human moral act is identified with the physical structure of the act. For the good of the person or the marriage itself, couples should be able to interfere with the sexual act [i.e., with its procreative potential]" (1995, p. 180).

For these theologians, presumably, moral evaluation of a human act includes reference to the intentions behind the act and to its likely effects, as well as to its nature or "physical structure." If in calculating the effects (combining all the relevant factors), some are negative while others are positive, it can be correct in this view to opt for the "lesser evil," and to do something that would otherwise be morally forbidden in order to avoid a greater evil (or achieve a greater good). In the same encyclopedia's entry on proportionalism, James Walter of Loyola University of Chicago claims that one of proportionalism's goals is to determine when one might (rightly) allow an exception to a moral norm. Proportionalists distinguish between moral evils (mainly what goes into an action–one's intentions and attitudes) and premoral evils (mainly the effects of the action–injury, pain, death, and the like). Exceptions to moral norms are granted when it is judged that the reasons for the act (its motives and intentions) combined with its good effects, are sufficiently valuable or important to outweigh the evil resulting from the act (pain and so on). Walter's example is the justification of killing in self-defense, since he takes the motive here (self-defense) to be good enough to outweigh the evil (the death of the attacker); hence he says this is an exception to the moral norm against killing.

Self-defense is a misleading example, however, since the Catholic moral tradition has always allowed acting in self-defense with whatever available means are necessary to stop the attacker. The exceptionless moral norm is not just "Never kill" but "Never intentionally take the life of an innocent human being." Hence, the example of self-defense does not fully disclose the nature of proportionalism. What

Walter omits from his account is that proportionalists also justify exceptions to norms that the tradition takes to be absolute. Consider an act of abortion in which the good intentions of the agent (perhaps a desire to finish college in four years and to raise a child only with the help of a spouse) and the premoral goods achieved (financial, emotional, and otherwise) might seem to outweigh both the moral evil of the act (the intention to take a life) and the premoral evils involved in it (the death of the baby). In such a case, presumably, proportionalist theologians could recommend choosing an abortion.

John Paul II defends the existence of moral absolutes and criticizes proportionalism (among other attacks on the tradition) in his encyclical *Veritatis Splendor* (The Splendor of Truth), citing the principle of Pope Paul VI in *Humanae Vitae* (1997, no. 14) that "it is never lawful, even for the gravest reasons, to do evil that good may come of it (cf. Rom 3:8)–in other words, to intend directly something which of its very nature contradicts the moral order, and which must therefore be judged unworthy of man, even though the [further or ultimate] intention is to protect or promote the welfare of an individual, of a family, or of society in general" (John Paul II, 1993, no. 80). Contrary to what Curran claims, traditional teaching on absolute norms does not identify a human act with the physical structure of the act, for on such a view, even unconscious or coerced 'actions' would be subject to moral appraisal, which is absurd. Rather, the tradition takes as morally determinative in a human act what the agent *wills to do*, which includes both the intention to perform this act and a description of the act that is intended. The centrality of the will has always been acknowledged within the tradition. One indication of this is that the moral appraisal of human acts has focused on what goes into them–the attitudes and intentions of the agent–rather than on their effects or consequences. Actions are judged morally according to the virtues or vices they express or indicate, be these natural virtues such as patience and justice or supernatural virtues like faith and love. "The morality of the human act depends primarily and fundamentally on the 'object' rationally chosen by the deliberate will. . . . The object of the act of willing is in fact a freely chosen kind of behavior. . . . Consequently, as the Catechism of the Catholic Church teaches, 'there are certain specific kinds of behavior that are always wrong to choose, because choosing them involves a disorder of the will, that is, a moral evil" (John Paul II, 1993, no. 78). The expected effects of one's actions will of course enter into one's deliberations about what would be good to do or to will, but what determines the morality or immorality of acts is what we *will*, not the effects of our actions in the world. Many of these effects will be outside of our control, and some will occur even after our lifetime, making an accurate calculation of them humanly impossible. Nor can additional intentions we might have, for example, to relieve someone's suffering, justify choosing to do something that violates his personal dignity (such as taking his life). In the personalist language of John Paul II, one is never morally justified in using another person or treating him or her as a means to an end, regardless of the independent desirability of that end. In every action directed toward persons, one must treat them as ends in themselves; hence, every such act must itself derive from an attitude of benevolence and love.

When we turn to the arena of marriage and sexuality, this approach rivets our attention on the fact that choosing a mate is choosing a *person*, which makes human coupling radically different from the mating practices of other animals. John Paul II explains:

> Clearly, if we are to speak of choosing a person, the value of the person must itself be the primary reason for the choice. . . The person as such must be the real object of choice, not values associated with that person, irrelevant to his or her intrinsic value. . . . So that although the sexual values in the object of choice may disappear, and however they may change, the fundamental value–that of the person–will remain. . . . The truth about the object of choice is attainable when, for the chooser, the value of the person as such is that to which all others are secondary. Sexual values, which act upon the senses and the emotions, are assigned their proper place (1981, p. 133).

The idea is not that one denies or suppresses physical and emotional responses to one's spouse, but that these are integrated with and subordinated to a deeper commitment to him or her–that is, to a genuine and selfless love. The Holy Father beautifully describes the mature love between a man and woman who have grown accustomed over the years to placing the person above all other values. In time, even the emotions and desires begin to conform to this deeper orientation of the will. "The love for a person which results from a [morally] valid act of choice is concentrated on the value of the person as such and makes us feel emotional love for the person as he or she really is, not for the person of our imagination, but for the real person. We love the person complete with all his or her virtues and faults, and up to a point independently of those virtues and in spite of those faults" (1981, p. 135).

Love based on utility or pleasure contains within itself the principle of its own disintegration. The love that we desire from our friends, and the friendship we desire from our spouses, must ground itself in something unshakeable–the inherent value of the person. To the extent that sex is divorced from love of this kind, it inevitably results in using a person. If one does not seek the good of the other person as such, the values that remain in the sexual act will generally be those that are subjective or centered on the self–pleasure, comfort, ecstasy, ego-enhancement, and the like. The other person may be an *occasion* for these values, but will not be the *end* of the sexual act. Some argue in favor of exceptions to absolute sexual norms in cases where, for example, a woman might have sexual relations with a man other than her husband (e.g., a prison guard) for the sake of her husband's welfare, so that he will be treated more kindly or perhaps even freed. Her intentions are not self-directed, then, but are directed toward the good of her husband. In this case, however, the woman commits an injustice toward the guard in using him sexually as a means to an end. She also does an injustice toward her husband to whom she owes sexual fidelity, not just a general commitment to his welfare.

Similar considerations apply to the case of a husband who seeks the good of his wife, the pleasure and emotional support involved in the sexual act, but who also wants to spare her the burden of another pregnancy, or to spare his family the financial strain another child would bring, or to avoid a risk of conceiving a physically handicapped child, and so on These are good intentions, but they do not remove the injustice involved in engaging in sexual acts that intentionally reject the potential fertility of the man or the woman, their potential fatherhood or motherhood. This potential is a good and not an evil; it is not a proper object of attack; further, it is intrinsic to the person as male or female and intrinsic to the meaning of the marital act itself. When there are serious reasons to avoid pregnancy, methods that time sexual intercourse around an awareness of the woman's times of fertility and infertility can be employed. Such methods of natural family planning can be practiced with morally wrong intentions too, of course, but they are not inherently depersonalizing in the way contraceptive sex is.[3]

3. IMPLICATIONS FOR SEXUAL MORALITY

Substantive conclusions about sexual morality follow fairly obviously from these reflections on acts and their meanings. The Holy Father develops an extensive analysis of the significance of the human body in the complementarity of masculinity and femininity and in the natural meaning of some of our bodily actions. This 'theology of the body' has roots at least as far back as St. Augustine, who speaks of "a kind of natural language common to all races which consists in facial expressions, glances of the eye, gestures, and the tones by which the voice expresses the mind's state" (Augustine, 1988, p. viii). The sexual act has a natural meaning too, grounded in its connection with procreation and the formation of the family, a meaning that is neither created by humans nor removable by their personal fiat. John Paul II sees this as a good thing, not as an unwelcome negation of our freedom to make our actions mean whatever we want them to mean. Because of the procreative meaning of the marriage act and the responsibilities attached to a marriage commitment and family life, couples can see their sexual relations as morally justified.

Sex is an arena rife with opportunities for exploitation; indeed some wonder whether treating the other (usually the woman) as an object can in principle be separated from sexual intercourse. Feminist writer Andrea Dworkin argues that even when a woman consents to sex, this only intensifies the damage done to her as a person.

> It is especially in the acceptance of the object status that her humanity is hurt: it is a metaphysical acceptance of lower status in sex and society; an implicit acceptance of less freedom, less privacy, less integrity. In becoming an object so that he can objectify her so that he can [expletive] her, she begins a political collaboration with his dominance; and then when he enters her, he confirms for himself and for her what she is: that she is something, not someone, certainly not someone equal (1987, pp. 140-141).

In a society pervaded by a contraceptive mentality where sexual love is rarely associated with marriage and children, there is a sting of truth in this indictment. Unfortunately, the temptation to use others for selfish ends can afflict every human person, including well-meaning Christian husbands and wives. But remaining open to the procreative potential of sex keeps it open to values that transcend the self–to children and responsibilities for them, to a partnership that goes beyond the moment and projects the couple into a shared future. As John Paul II puts it, "When a man and a woman who have marital intercourse decisively preclude the possibility of paternity and maternity, their intentions are thereby diverted from the person and directed to mere enjoyment: 'the person as co-creator of love' disappears and there remains only the 'partner in an erotic experience'" (1981, p. 234). In each act of betrothed love, the focus must be on the other person, filled with love for them and a desire for their good. Otherwise this act will in some way objectify the other person, violating his or her inherent dignity.

Sexual acts outside of marriage clearly cannot be justified within an ethic of respect for the value of persons as such. Extramarital sex cannot express a total and unconditional commitment to the good of one's sex partner, since there is no such commitment to him or her–the person as such is not the end of these acts; some form of enjoyment (or another end independent of the value of that person) is the end. One of Wojtyla's important new insights here is his claim that even within marriage, an act of sexual intercourse may be immoral. It violates the dignity of the other party whenever it is motivated solely by selfish interests rather than by love. It's not that one cannot desire physical and emotional satisfaction as well, but these should be subordinated to or at the service of love of the person for his or her own sake. "Adultery in the heart is committed not only because man looks in this [lustful] way at a woman who is not his wife, but *precisely* because he looks at a woman in this way. Even if he looked in this way at the woman who is his wife, he could likewise commit adultery in his heart" (John Paul II, 1997, p. 157). Women are perhaps more sensitive to the possibilities for abuse even within marriage, and the Pope suggests that husbands should take special care to see that sexual expressions of love affirm their wives *as persons.*

Genuine love for one's spouse requires openness to procreation in sexual relations. Human persons are sexual beings, and the body and its powers are part of their glory. To love a human person, to truly desire his or her good, requires an acceptance of the body as well–its natural gifts and capacities. Here again, though, the person cannot be treated merely as a means for the end of procreation, but must be loved for his or her own sake. The norm of love does not require that couples intend in each sexual act to conceive a child, but that this intrinsic power proper to the other person in his or her body is not deliberately opposed or negated. Sex has many meanings, and its primary meaning is to express marital love, 'a total gift of oneself to the other' in John Paul II's words. This is why married couples who know themselves to be infertile, whether temporarily or permanently, do something good and appropriate in expressing their love for each other sexually. "Infertility in itself is not incompatible with inner willingness to accept conception, should it occur" (1981, p. 236). The willingness is the key, not the result of that willingness.

Some theologians argue that recourse to artificial means of preventing conception can sometimes be justified in cases where a married couple has a general willingness to accept children (at some time or other), and when the marital act is motivated in part at least by morally good intentions–for example, to give pleasure to each other, to express love, to console or encourage each other and so on. But this rationale assumes that a contracepted sexual act, when freely chosen, is not in itself (as an intended act) unjust or unloving. If the act is otherwise morally good or neutral, then of course it is morally justifiable in cases where no vicious intentional states enter into it; but if it is a morally objectionable act, then further motives and ends cannot serve to justify it. We can see this more clearly perhaps in the case of acts that everyone recognizes as unjust, such as date rape (coercing a person into sexual intercourse without her or his consent). Whatever 'loving' purposes may lie behind someone's choice of this act, the decision to commit *this act* is unloving. *One cannot perform an unloving act lovingly*, though one may see it as a *means* to some further morally benign *end.* Such moral reasoning shows its true colors, as a form of utilitarianism, and hence a rejection of the supreme value and inviolability of persons.

4. REPRODUCTIVE TECHNOLOGIES

Many public statements from the Vatican in the last two decades or more on artificial methods for manufacturing babies draw on the personalist norm in rejecting those techniques that undermine the dignity of the human being–especially the newly-formed human lives that result from these techniques. The Vatican statement *Donum Vitae (Instruction on Respect for Human Life in its Origins and for the Dignity of Procreation)* issued in 1987 by Joseph Cardinal Ratzinger as Prefect of the Congregation for the Doctrine of the Faith, cites several moral principles drawn from previous documents of the Church, especially those authored by John Paul II. It begins from the premise that "science and technology are valuable resources for man," but "being ordered to man, who initiates and develops them, they draw from the person and his moral values the indication of their purpose and the awareness of their limits" (Congregation for the Doctrine of the Faith, 1987, Intro. no. 2). This introductory section of the instruction also draws attention to the dignity and importance of the body as "a constitutive part of the person who manifests and expresses himself through it" (CDF, 1987, I, no. 2).

Moral objections arise here to both *in vitro* fertilization and artificial insemination, whether the egg and sperm come from a third party or from the spouses themselves, and whether or not some embryos are deliberately destroyed in the process. The objections come from two considerations: the dignity of the child and the dignity of the spouses in their sexual union, i.e, their (in principle) procreative acts. The inviolability of the bodily life of a human being stems from the inviolability of the person whose life is a bodily one. The duty to respect and protect the life of every human being, from the moment of conception until natural death, applies equally to human beings at every stage of development–it attaches to them qua human, in other words, and not in virtue of other qualities or attributes they may have. Lest there be any confusion, the document explicitly proclaims that every individual human being is a human person. Attempts to dissociate these two concepts have ever been in the service of eliminating the unwanted or the 'unworthy of life.'

Drawing on these principles, it follows immediately that any intentional harm or destruction directed toward human embryos is morally reprehensible, as a frontal assault on human dignity of the first order. Use of such beginning humans for experimentation or other purposes likewise treats them as mere means to ends outside themselves, and so cannot be made licit either by the desirability of these further ends or by the 'consent' of the parents or 'donors.' No one may give consent on behalf of another person to accept an immoral using of them. But not all reproductive techniques require that embryos be destroyed or indefinitely 'frozen'. If a husband's sperm is 'harvested' for transplantation into the wife's fallopian tubes, for example, no intentional destruction of the resulting embryo need be involved. However, even these scientific techniques fail to conform to the respect that is owed to all the parties involved.

In the first place, the document argues that an injustice is done to the human being produced in this way, for the very reason that he or she becomes a product. "It is through the secure and recognized relationship to his own parents that the child can discover his identity and achieve his own proper human development" (CDF, 1987, II, no. 1). Children naturally conceived are meant to be the fruit of marital love–an occasion for joy, received as a gift to their parents, a living sign of their love and union. Although the injustice is more obvious in cases where the child is not genetically tied to both parents, it is also true of the child conceived by technical processes distinct from the marriage act. A child "cannot be desired or conceived as the product of an intervention of medical or biological techniques; that would be equivalent to reducing him to an object of scientific technology" (CDF, 1987, II, no. 5) wherein his coming-to-be is evaluated by standards of technical efficiency and is under the direct control of other persons.

It is worth noting that this focus on the good of the newly-conceived child has implications for the moral gravity of other actions as well. Extramarital sex, for example, is an offense not just to the dignity of the parties involved, but to the dignity of any child resulting from such acts. The question is not whether that child's life has value independently of the way in which it was conceived–of course the answer is resoundingly in the affirmative. Rather, the question is whether it is an injustice to the child to have risked her sense of self, her sense of belonging and being loved, by engaging in actions that could (and often do) lead to a child's coming-to-be (a great good, to be sure) without a family to which she belongs (a great harm).

A further moral principle relevant to reproductive technologies is the importance of the dignity proper to married couples and to their acts of sexual love. "The bond existing between husband and wife accords the spouses, in an objective and inalienable manner, the exclusive right to become father and mother solely through each other" (CDF, 1987, II.A, no. 2). It offends against the dignity of one's spouse, then, to become a parent by way of the genetic contributions of a third party. Even when the spouses themselves provide the 'genetic materials,' however, they become parents through a scientific process rather than through the intimacy of their acts of love. This dissociates the procreative dimension of these acts from their meaning as expressions of love of the other person as such. While this problem is touched on quite briefly in *Donum Vitae*, it was developed earlier by Wojtyla in *Love and Responsibility*. "Marital intercourse is in itself an interpersonal act, an act of betrothed love, so that the intentions and the attention of each partner must be fixed upon the other, upon his or her true good. They must not be concentrated on the possible consequences of the act, especially if that would mean a diversion of attention from the partner" (John Paul II, 1981, pp. 233-234). Although the focus in this passage is the spouses' intentions in the act of intercourse rather than on their intentions in using artificial techniques (along with intercourse perhaps) for conceiving a child, the moral principle involved is the same. Such intentions offend the dignity of the spouses as well as the dignity of the resulting child. They divert the spouses' intentions (and attention) in some sexual acts from the beloved and toward a different end.

The moral reasoning behind these objections to IVF and its cousins applies equally to human cloning. The truth is that most attempts to justify cloning (whether for therapeutic or reproductive purposes) would justify embryonic experimentation and *in vitro* fertilization as well. Michael Sandel, professor of government at Harvard and a current member of the President's Council on Biomedical Ethics, while he disagrees with an absolute ban on cloning, clearly recognizes the logical connections between cloning for biomedical research and current IVF practices that produce several embryos with no intention of preserving all of them. He writes:

> Those who oppose the creation of embryos for stem cell research but support research on embryos left over from *in vitro* fertilization (IVF) clinics beg the question whether those IVF 'spares' should have been created in the first place: if it is immoral to create and sacrifice embryos for the sake of curing or treating devastating diseases, why isn't it also objectionable to create and discard spare IVF embryos for the sake of treating infertility? After all, both practices serve worthy ends, and curing diseases such as Parkinson's, Alzheimer's, and diabetes is at least as important as enabling infertile couples to have genetically related children (2002, pp. 343-344).

Sandel relies on consequence-driven reasoning here, whatever his wider moral theory would hold. He describes the utilitarian reasoning behind our current justification of IVF procedures, and points out that an exactly similar justification could easily be provided for cloning. The moral justification of IVF in the United States today, he says, "rests on the idea that the good achieved outweighs the loss." This cost/benefit analysis applies in such instances only if Sandel is right in thinking that "the loss is not of a kind that violates the respect embryos are due" (Sandel, 2002, p. 344).

Sandel presented a more extended explication and defense of his view in a recent lecture at Harvard's Kennedy School of Government entitled "The Ethics of Human Cloning" (delivered on November 18, 2002).[4] He characterizes his position as the reasonable mean between the extremes, since he opts neither for the view that embryos deserve the same respect that a grown person deserves nor for the view that embryos are mere things deserving of no respect. He thinks them deserving of some intermediate level of respect that does not necessarily rule out using them for serious enough purposes. In his published statement to the President's Commission, Sandel tells us that we should not carve our initials in an ancient sequoia tree, he tells us, "not because we regard the sequoia as a person but because we consider it a natural wonder worthy of appreciation and awe" (Sandel, 2002, p. 345). Unfortunately for the embryo, "respecting the forest may be consistent with using it. But the purposes must be weighty and appropriate to the wondrous nature of the thing" (2002, p. 345).

What this line of reasoning illustrates is the intrinsic connection between personhood and inviolability. Whatever is not a person, whatever is less than a person, we may sometimes be justified in using as a means to a (sufficiently significant) end. To take one especially disturbing example, a recent article by J. Bottum in *The Weekly Standard* reminds us that "a few years ago, the *London Daily Telegraph* reported that 'doctors at the state-run Shenzhen Health Centre for Women and Children hand out bottles of thumb-sized aborted babies to be made into meat cakes or soup with pork and ginger. Zou Qin, a doctor at the Luo Hu Clinic in Shenzhen, said the fetuses were 'nutritious' and that she had eaten one hundred herself in the last six months. 'We don't carry out abortions just to eat fetuses,' said Qin, '[But they would be] wasted if not eaten'" (Bottum, 2003).

Bottum describes a recent documentary on British television about Chinese 'performance artist' Zhu Yu "who displays photographs in which he washes a stillborn child in a sink and then consumes it" (Bottum, 2003). When a culture is as deeply committed as ours is (by permissive abortion laws and IVF technologies) to the valuelessness of new human life, it is difficult indeed to find a convincing moral objection to what the executives at Channel 4 found merely 'thought-provoking'.

This does nothing by itself to decide the moral status of the human embryo, but many members of the President's Council, among others, provide excellent grounds, philosophical and scientific, for regarding embryos as tiny human beings and hence as human persons. The appendix to the council's report contains personal statements by several individual members, including an especially insightful contribution by Professors Robert George of Princeton University and Alfonso Gómez-Lobo of Georgetown University spelling out the case for respect for even these newly-conceived human beings (George and Gómez-Lobo, 2002, pp. 294-306).

The current debate and division in our society about the permissibility of human cloning stems from a deeper division about the basis of morality, one that can be broadly characterized as a dispute between consequentialism and non-consequentialism. A consequentialist moral theory is one in which moral justification of an act (or a policy or program) stems solely from its results, direct and indirect. Non-consequentialists allow that other factors (especially, whether the act is intended, the spirit in which it is done, the circumstances, one's past commitments, and the like) can determine the rightness or wrongness of the act (or program or policy) . Non-consequentialist theories include the kind of virtue ethic endorsed by Aristotle and St. Thomas Aquinas as well as 'deontological' theories endorsed by Immanuel Kant and W. D. Ross. Included in the non-consequentialist camp is John Paul II's personalism, which places respect for the dignity of persons at the core of morality and insists that every act directed toward persons must acknowledge their value. They must be loving acts with respect to that person (not merely with respect to other persons or goods).

To the extent that consequentialist thinking dominates the moral consciousness of our nation (as well as the academy), we will find it impossible to provide credible arguments against the rapid advance of dehumanizing technologies and policies. Arguments that appeal to utility are never finished; they are open to revision whenever a different outcome is predicted. Professor Rebecca Dresser of Washington University is a former member of the President's Council on Bioethics with expertise in both law and medical ethics. In her personal statement on the cloning debate, "a central ethical issue [and the only one she raises] is whether studies of cloning to have a child would present a balance of risks . . . and expected benefits . . . that justifies proceeding with human trials" (Dresser, 2002, p. 283). Although she admits that "to approve cloning is to approve the creation of embryos as research tools," she continues with barely a blink of the eye, "I can imagine studies that would offer sufficient benefit to patients to justify the creation of embryos for research through cloning or other methods" (2002, p. 286).

Among the greatest contributions of Karol Wojtyla, before and after his election to the papacy, is his constant insistence on the primacy of the person–both the transcendence and inherent glory of persons and the way in which that glory transfigures even those aspects of human nature that are shared with other living things. In what might be his most widely-cited work, the encyclical *Evangelium Vitae* (The Gospel of Life), John Paul II quotes in its entirety the story of Cain and Abel from the Book of Genesis. Cain has murdered his brother Abel out of anger and envy, and when questioned by God about the whereabouts of Abel, he blurts out: "I do not know. Am I my brother's keeper?" (Gn 4:9).

> "*I do not know*": Cain tries to cover up his crime with a lie. This was and still is the case when all kinds of ideologies try to justify and disguise the most atrocious crimes against human beings. "*Am I my brother's keeper?*": Cain does not wish to think about his brother and refuses to accept the responsibility which every person has toward others. We cannot but think of today's tendency for people to refuse to accept responsibility for their brothers and sisters (1995, no. 8).

Professor Alta Charo is Associate Dean of the University of Wisconsin Law School and a former member of the National Bioethics Advisory Council that issued a report on human cloning in 1997. In an interview on 'Breaking News' with CNN anchor Fredricka Whitfield (January 4, 2003), Charo fretted over the possibility that there might be a move to ban all human cloning and research on embryos based on what she believes is the irresponsible behavior of those who claim to be cloning for reproductive purposes. "We've already seen members of Congress moved to write editorials in the 'USA Today' or make statements to the press they plan to press forward with legislation that goes way beyond anything we need with regard to reproductive cloning and move toward banning important medical research that uses some of the same techniques with no reproductive outcomes" (Charo, 2003). Of course, if the human embryo is already such an 'outcome' (i.e., a child), then producing him or her as a research tool is morally repugnant, not just irresponsible. But what is also troubling in Charo's remarks is that the very language used betrays a nonchalant attitude toward human life. If a human child can be described as a 'reproductive outcome', there is little to prevent us from placing him or her at the service of other ends. Am I my 'outcome's' keeper? Surely not.

The richness of the moral reflections of Pope John Paul II continues to attract the attention of scholars, however, and we may hope that it will ultimately serve as a wake-up call for a society steeped in utilitarian thinking. "When conscience, this bright lamp of the soul (cf. Mt 6:22-23), calls 'evil good and good evil' (Is 5:20), it is already on the path to the most alarming corruption and the darkest moral blindness" (John Paul II, 1995, no. 24). Fortunately, the Holy Father believes in the power of One who can make the deaf hear and the blind see, and who continues to speak to the heart of every person. "The Church puts herself always and only at the *service of conscience*, helping it to avoid being tossed to and fro by every wind of doctrine proposed by human deceit (cf. Eph 4:14), and helping it not to swerve from the truth about the good of man, but rather, especially in more difficult questions, to attain the truth with certainty and to abide in it" (1993, no. 64).

Boston College
Chestnut Hill, Massachusetts

NOTES

1 The same principle cannot be applied to animals or other natural beings, since these do not choose their own ends. Moral obligations accompany the use of natural resources and interactions with animals, but these have a separate basis–one of respect appropriate to this kind of being; one might also take into consideration the indirect effects of these actions on other persons (present and future).

2 This discussion occurs in one of Pope John Paul II's Wednesday audiences, 'Original Innocence and Man's Historical State,' published in *The Theology of the Human Body: Human Love in the Divine Plan,* (pp. 72-74).

3 The focus here is on the objective rightness or wrongness of actions, not on the degree of culpability in the agent. This culpability can be lessened by various forms of coercion or invincible ignorance, but these are outside the scope of this paper.

4 A summary of Sandel's lecture is available at the website for Harvard University's Center for Ethics and the Professions: http://www.ethics.harvard.edu/summaries/sandel.doc.

REFERENCES

Augustine (1988). *Confessions* F.J. Sheed (Trans.), New York: Rowman and Littlefield.

Bottum, J. (2003). 'Eating Babies: A Modest Proposal in China Shows Where the Brave-New-World Crowd Has Brought Us,' in *The Weekly Standard* [On-line], January 3, 2003. Available: http://www.weeklystandard.com/Utilities/printer_preview.asp?idArticle=2064&R=7598276.

Charo, A. (2003). Remarks to anchor F. Whitfield in 'Raelians Claim Second Cloned Baby Born,' CNN *Breaking News*, aired January 4, 2003. Available: http://www.cnn.com/TRANSCRIPTS/0301/04/bn.01.html.

Congregation for the Doctrine of the Faith (1987). *Respect for Human Life (Donum Vitae)*. Boston: Pauline Books and Media.

Curran, C. (1995). 'Birth Control,' in R. McBrien (Ed.), *The HarperCollins Encyclopedia of Catholicism*. San Francisco: HarperCollins. 179-181.

Dresser, R. (2002). 'Statement of Professor Dresser,' in *Human Cloning and Human Dignity: The Report of the President's Council on Bioethics*. New York: Public Affairs. 283-287.

Dworkin, A. (1987). *Intercourse*. New York: Free Press.

George, R. & A. Gómez-Lobo (2002). 'Statement of Professor George (Joined by Dr. Gómez-Lobo),' in *Human Cloning and Human Dignity: The Report of the President's Council on Bioethics*. New York: Public Affairs. 294-306.

John Paul II (Wojtyla, K.). (1981). *Love and Responsibility* [first published in 1960], H. Willetts (Trans.), New York: Farrar, Straus and Giroux.

John Paul II (1993). *The Splendor of Truth (Veritatis Splendor)*. Boston: Pauline Books and Media.

John Paul II (1997). 'Original Unity of Man and Woman: Catechesis on the Book of Genesis,' [first published in 1980] in *The Theology of the Body: Human Love in the Divine Plan*. Boston: Pauline Books and Media. 25-102.

John Paul II (1997). *Evangelium Vitae* (*The Gospel of Life*). [first published in 1995] in *The Theology of the Body: Human Love in the Divine Plan*. Boston: Pauline Books and Media. 427-442.

Paul VI (1997). *Humanae Vitae* (*Of Human Life*). [first published in 1968] in *The Theology of the Body: Human Love in the Divine Plan*. Boston: Pauline Books and Media. 493-582.

Sandel, M. (2002). 'Statement of Professor Sandel,' in *Human Cloning and Human Dignity: The Report of the President's Council on Bioethics.*,New York: PublicAffairs. 343-347.

CHAPTER SEVEN

PATRICK LEE

THE HUMAN BODY AND SEXUALITY IN THE TEACHING OF POPE JOHN PAUL II

1. INTRODUCTION

It is well known that at times some Christian thinkers have adopted implicitly a very dualistic view of the human person – dualistic in the sense of viewing the self as something which *has* or *inhabits* a body, rather than being a living, bodily entity. For example, some have thought of heaven, or the completed kingdom of God, as something purely spiritual, with no thought given to the role of the body.[1] On that view, bodily goods, including biological life, turn out to be nothing but mere means, or perhaps temporary vessels, for what alone is intrinsically valuable (*Gaudium et Spes*, no. 34). Recently some Christian thinkers have adopted a basically Kantian view of human action, downgrading conscious choices bearing on limited goods–"categorial choices"–as quite secondary or peripheral to a mysterious, sub-conscious "transcendental," fundamental option toward or away from the Absolute, God (Rahner, 1986, pp. 24-44, 90-106). On that view, once again, bodily goods, being the objects of mere "categorial choices," are demoted to the level of mere external signs of what is truly important, and purely spiritual. Such views lend credence to the widely held secular view of Christianity that it has a dim view of the human body and matter.

2. THE PROBLEM OF OBJECTIFICATION

The ethical teaching of Pope John Paul II, especially that concerning sexuality and marriage, should help to show that such dualism is simply a distortion of Christian teaching. That the human person is essentially a bodily being, a unity of body and soul, and that therefore the masculinity or femininity of the human being is internal to his or her personhood (rather than just interesting external "equipment"), has been a constant theme of Pope John Paul's teaching throughout his pontificate. In this paper I would like to explore the logical consequences for the basic ethics of sexuality and marriage of Pope John Paul II's insistence on the unity of the body and the soul. Specifically, I would like to suggest a particular interpretation, or rendering, of his "theology of the body" as it applies to the core issue of sexual ethics: under what conditions are sexual acts morally right? This issue is obviously a central one in bioethics, since most of the questions regarding fertility, reproductive technology, and sexual identity depend in some way on an answer to

C. Tollefsen (Ed.), John Paul II's Contribution to Catholic Bioethics, pp. 107–120.

that question. A perennial question in sexual ethics is: is there something "special" about sex? That is, are there moral norms which govern sex in addition to the prohibitions of deception and coercion?[2] It seems that there is something special about sex, and it seems that we can be aware of this point whether we accept revelation or not. For example, it seems clear to most people that a punch in the nose is far less serious than rape, although both involve violence. And it seems that this can be true only if sexual acts have some feature or features making them significantly different from other bodily acts.

Scott Anderson points out that, while some would argue that commercial sex (prostitution) is actually not ethically wrong, and should be legalized and normalized, very few would be willing to accept the concrete consequences of this (Anderson, 2002). Compare the following two scenarios. In the first scenario, A buys B an expensive dinner with the mutual, though tacit, understanding that B will tutor A in philosophy to prepare for A's upcoming mid-term exam. B accepts the dinner but then wishes to back out of the deal. In the second scenario, C buys D an expensive dinner with the mutual, though tacit, understanding that D will have sex with C. D accepts the dinner but then wishes to back out of the deal. I don't think we would hesitate to say that in the first scenario B owes it to A to follow through with the deal. Moreover, if the agreement had been formalized by a signed contract we would not hesitate to say that civil authorities could with justice enforce the terms of their agreement. However, our attitude is surely different with respect to the second scenario. As Anderson points out, although we may think it bad for D to agree to such a bargain, still: "we don't hold that the frustrated party is entitled to enforcement of the bargain against the wishes of his date" (2002, p. 775).

In other words, it is clear that sex is quite different from other bodily acts. A and B could enter a contract involving physical therapy, dental work, hair cutting, and so on, all actions which B might perform on A's body. And with respect to such acts we have no hesitation about their being commercialized, and the relevant contracts being enforced. A may hire B wholly or in part because B is a skilful physical therapist. Moreover, B's continuing to provide that service may quite properly be a condition of his or her continued employment. But most of us would at least balk at saying that C could hire D wholly or in part because of her sexual skills and make her continued employment conditional upon her continuing to provide those services (Anderson, 2002).

Moreover, we do not hesitate to say that bodily actions such as physical therapy are appropriate actions to perform with anyone, including one's parents, one's children, or even lower animals (as in veterinary). Yet sexual actions are not appropriate with our children or parents, or with other animals. Thus, there must be some feature or features of sex which do make it quite distinct from other bodily actions. But what is it? What is it about sex that makes it quite different from other bodily acts?

In sexuality more than in other areas we tend to describe immoral instances of it as treating someone as a mere thing, as objectification. Is there some feature or features whose presence means that the act is an immoral act of manipulation? What must a sexual act possess in order *not* to be an instance of immoral manipulation?

Thomas Mappes argues that voluntary and informed consent to an act is sufficient to render it different from immoral manipulation (Mappes, 2002). But if this were true, then our different reactions to the different scenarios described above would be groundless. If he were correct, then the woman who enjoyed dinner at the lavish spender's expense not only would be engaging in deceit (indeed we might be tempted to say that she too, though in a different way, was merely using him), but she would *owe* it to him to fulfill her part of the tacit bargain.[3] This, however, seems to be incorrect. Mappes's position cannot account for the different attitudes we have toward sexual acts and other bodily acts.

Moreover, I believe some feminists have correctly shown that pornography involves an attitude toward women as mere things or toys for use[4] even though the participants may have informed consent. Martha Nussbaum commented on the case that Andrea Dworkin and others have made regarding pornography. Nussbaum evidently does not agree that *all* pornography is reductive, but she does hold that *Playboy* magazine involves treating persons as mere things. Her appraisal is worth noting: "*Playboy* depicts a thoroughgoing fungibility and commodification of sex partners, and, in the process, severs sex from any deep connection with self-expression or emotion (Nussbaum, 2002, p. 406)." This seems correct. It seems that to avoid reducing persons to mere things, a sexual act must in some way be connected to "self-expression or emotion." But the question remains: what must be the case for a sexual act to have this connection?

3. LANGUAGE OF THE BODY AND SEXUAL MORALITY

Pope John Paul teaches that in order for a sexual act to be non-objectifying it must express a marital communion. In that way it realizes a basic human good, the marital communion of the spouses, and thus does treat both oneself and the other as subjects, rather than as mere objects of use. This, of course, is a traditional teaching of the Catholic Church. Yet John Paul's explanation of this teaching is quite distinctive; it differs from the basic arguments usually attributed to friends of traditional sexual morality. For it is often assumed in treatments of sexual ethics that the central argument in its defense is simply that extra-marital acts, contraception, etc., are unnatural, that is, contrary to the direction inscribed in the reproductive or procreative power. This argument, often described as the "physicalist" or "naturalist" argument, has been, in my judgment, rightly criticized.[5] For example, it is not clear that acting contrary to the orientation of a biological power, or contrary to the natural orientation of the sexual act, is necessarily wrong. Nor is it clear that all extramarital sexual acts are really contrary to that direction (instead of being outside it.) Pope John Paul himself has rejected this kind of argument:

> Therefore this law [that is, natural moral law] cannot be thought of as simply a set of norms on the biological level; rather it must be defined as the rational order whereby man is called by the Creator to direct and regulate his life and actions and in particular to make use of his own body. To give an example, the origin and the foundation of the duty of absolute respect for human life are to be found in the dignity proper to the person and not simply in the natural inclination to preserve one's own physical life. Human life, even though it is a fundamental good of man, thus acquires a moral

> significance in reference to the good of the person, who must always be affirmed for his own sake (1993, no. 50).

It is true that, speaking as the visible leader of the Catholic Church, the Pope is not expected to provide an extensive *philosophical* defense of his teachings. After all, the claim of the popes, including in this century Pius XI, Pius XII, John XXIII, Paul VI, as well as John Paul II, has been that the Catholic Church's basic teachings on sexuality and marriage are part of the teaching handed down from the Apostles and divinely guaranteed.[6] Still, some *explanation* is usually provided, and philosophical arguments can be extracted from such explanations.

Often John Paul explains why only within marriage is sexual intercourse non-objectifying by referring to what he calls "the language of the body." For example, in his commentary on the book of *Genesis* in his weekly audiences in the beginning of his pontificate, he describes the sexual act as a kind of "prophetism of the body" (1997, p. 358). And he teaches that sexual intercourse is a language that has an objective meaning, and that its meaning is full conjugal communion:

> In the texts of the prophets the human body speaks a "language" which it is not the author of. Its author is man as male or female, as husband or wife – man with his everlasting vocation to the communion of persons (1997, p. 359).

He then argues that to have sexual intercourse with someone who is not one's spouse is a type of lie. Fornication and adultery contradict the language objectively inscribed within man and woman and their two-in-one-flesh union:

> If the texts of the prophets indicate conjugal fidelity and chastity as "truth," and adultery or harlotry, on the other hand, as "non-truth," as a falsity of the language of the body, this happens because in the first case the subject (that is, Israel as a spouse) is in accord with the spousal significance which corresponds to the human body (because of its masculinity or femininity) in the integral structure of the person. In the second case, however, the same subject contradicts and opposes this significance (1997, p. 360).

Such passages can be interpreted in two different ways. At first glance one might interpret such passages as arguing as follows: It belongs to the nature of the sexual act to signify total self-giving. Yet to engage in the act and withhold an aspect of the self is to contradict the natural signification of this act. And to contradict the natural signification of the sexual act is wrong.

There are difficulties with this argument. In fact this argument is not, in its structure, very different from the simple "physicalist" or "naturalist" argument mentioned above, and as a consequence it has similar logical defects. Even granted that it belongs to the nature of the sexual act to express total personal communion, why should it be in itself wrong to alter the natural symbolism of this act? If it is not in itself wrong to act contrary to the nature of the natural teleology of an act or power, why would it be in itself wrong to act against, or not fully in accord with, the natural symbolism of an act? Or one might instead question whether sexual intercourse necessarily, or by its nature, symbolizes *total* personal communion. Some might argue that it is an apt symbol for a close personal, though not necessarily permanent, union – a union not necessarily open to having children, and not necessarily heterosexual. In short, if the sexual act is viewed fundamentally as a *symbol* – that is, if one views *symbolizing* as the basic type of act one is performing when one has sex – then it is not clear that its natural meaning is (morally) inalterable or that its meaning is necessarily what the Pope has claimed it is (namely, *marital communion*).

4. POPE JOHN PAUL ON BODY AND SOUL

However, there is a second, and I believe more accurate, way of interpreting the argument referring to sexual acts as "language of the body." The rest of this article will be devoted to explaining this second interpretation. To understand John Paul's teaching we must return to his re-affirmation of the unity of the body and the soul in the human being. John Paul insists at several points in clear terms that we *are* bodily entities, that we are living bodies – rational animals and persons, but essentially bodies at the same time. For example, John Paul does not hesitate to say that human persons are bodies, albeit, he adds, also having self-consciousness, free choice, and immortal souls:

> However, the fact that man is a "body" belongs to the structure of the personal subject more deeply than the fact that in his somatic constitution he is also male or female (1997, p. 43).

> As the expression of self-determination, choice rests on the foundation of his [man's] self-consciousness. Only on the basis of the structure peculiar to man is he a "body" and, through the body, also male and female (1997, p. 50).

> Thus formed [that is, created in God's image], man belongs to the visible world; he is a body among bodies (1997, p. 38).

Later in this same work, John Paul comments on the *Genesis* text which says: "So God created man in his own image, in the image of God he created him; male and female he created them," (1:27). The standard interpretation of this text has usually been that man is the image of God in his mind, in his intellect and will. Without denying this point, John Paul boldly teaches that man and woman are created in God's image precisely in their body, precisely in their complementarity, which orients them to full marital communion.

> The same "man," as male and female, knowing each other in this specific community-communion of persons, in which they are united so closely with each other as to become "one flesh," constitutes humanity. That is, they confirm and renew the existence of man as the image of God (1997, p. 83).[7]

Moreover, John Paul insists at several points that the truth that human persons are body and soul, not just souls, and not just consciousnesses, is central for ethical issues, especially in sexual ethics. In his encyclical *Veritatis Splendor* (*The Splendor of Truth*), he replies to ethical theories which would reduce the biological aspects of the human person to mere presuppositions for action and thus deny their intrinsic importance. To such views he replies as follows:

> This moral theory does not correspond to the truth about man and his freedom. It contradicts the *Church's teachings on the unity of the human person,* whose rational soul is *per se et essentialiter* the form of his body [here a footnote refers to the Ecumenical Council of Vienne and The Fifth Lateran Ecumenical Council]. The spiritual and immortal soul is the principle of unity of the human being, whereby it exists as a whole – *corpore et anima unus* as a person [here a footnote refers to the

> Second Vatican Council, *Gaudium et Spes, no. 14*]. These definitions not only point out that the body, which has been promised the resurrection, will also share in glory. They also remind us that reason and free will are linked with all the bodily and sense faculties (1993, no. 48).

And in the Apostolic Exhortation *Familiaris Consortio* (*On the Christian Family*) he explains that sexual immorality involves in some way an alienation of the body from the spirit. He teaches that only marital sexual acts that express marital communion and are open to procreation, "recognize both the spiritual and corporal character of conjugal communion" (1981, no. 32). He then says: "In this way sexuality is respected and promoted and in its truly and fully human dimension, and is never "used" as an "object" that, by breaking the personal unity of soul and body, strikes at God's creation itself at the level of the deepest interaction of nature and person (1993, no. 48)."

5. SEXUAL INTERCOURSE AS SUBSTRATUM AND CONSTITUENT OF MARRIAGE

This doctrine of the unity of the body and the soul is important for understanding John Paul's teaching on sexual morality because, according to that teaching, the sexual act is *not* a mere extrinsic sign or symbol (as the misinterpretation of the "Language of the Body" argument would have it), but *expresses* or *embodies* a personal communion.

To see how, we must first note John Paul's teachings on the complementarity of man and woman, and on marriage, which is the personal communion to which the differentiation of the sexes and their complementarity are intrinsically oriented. Man and woman are complementary. In a sense human nature is complete only in the two together. Their union, on the level of the body, emotions, and spirit, is marriage. And this union is a fundamental human good, that is, intrinsically good as opposed to merely instrumentally good:

> This conjugal communion sinks its roots in the natural complementarity that exists between man and woman, and is nurtured through the personal willingness of the spouses to share their entire life-project, what they have and what they are: for this reason such communion is the fruit and the sign of a profoundly human need (1981, no. 19).

Thus, marriage is the fulfillment of the basic need or natural inclination toward complementary personal communion. It is the union of man and woman precisely as man and woman.

Commenting on the Genesis text, "The man and the wife were both naked, and were not ashamed" (2:25), John Paul speaks of an original innocence in which the man and the woman viewed each other as bodily persons. And they understood their masculinity and femininity *as intrinsically orienting them to personal communion, or reciprocal self-gift, in marriage*. Without a return, in a certain sense, to this innocence and an *acceptance* of the other as gift, the sexual act is reduced to a mere *use* of the other for one's own satisfaction:

> This dignity corresponds profoundly to the fact that the Creator willed (and continually wills) man, male and female, "for his own sake." The innocence "of the heart," and

> consequently the innocence of the experience, means a moral participation in the eternal and permanent act of God's will. The opposite of this "welcoming" or "acceptance" of the other human being as a gift would be a privation of the gift itself. Therefore, it would be a changing and even a reduction of the other to an "object for myself" (an object of lust, of misappropriation, etc.) (1997, p. 70).

It is a multi-leveled union. That is, it is essentially bodily as well as spiritual:

> The gift of the Spirit is a commandment of life for Christian spouses and at the same time a stimulating impulse so that every day they may progress towards an ever richer union with each other on all levels – of the body, of the character, of the heart, of the intelligence and will, of the soul – revealing in this way to the Church and to the world the new communion of love, given by the grace of Christ (1997, p. 70).

This point is distinctive and important. Many theologians and philosophers have viewed marriage as essentially a spiritual reality, a distinctive sort of friendship. They have then viewed the sexual act within marriage as an extrinsic sign that fosters marital love and friendship by signifying it. In effect, the inaccurate interpretation of the "language of the body" argument (which also makes it unsound) discussed above presupposes this view of marriage and of the sexual act's relationship to marriage. John Paul's teaching is different. According to John Paul, marriage *includes* bodily union.

This position, he points out, is implied by the constant teaching of the Church (though the implication has often been missed), that marriage is not *consummated* until there is sexual intercourse.[8] As a consequence, the way the sexual act contributes to marriage, according to the teaching of John Paul, is by embodying it, expressing it, making it bodily present. It is, he says, the bodily *constituent or substratum* of that multi-leveled union:

> The consummation of marriage, the specific *consummatum, is also enclosed in this knowledge* [the sexual intercourse which the bible refers to as "knowing"]. In this way the reaching of the "objectivity" of the body, hidden in the somatic potentialities of the man and of the woman, is obtained, and at the same time the reaching of the objectivity of the man who "is" this body. By means of the body, the human person is husband and wife. At the same time, in this particular act of knowledge, mediated by personal femininity and masculinity, the discovery of the pure subjectivity of the gift – that is, mutual self-fulfillment in the gift – seems to be reached (1997, p. 81).[9]

Thus, morally right sexual intercourse in marriage is not merely instrumental to other goods. It is not just a means of giving and obtaining pleasure. Nor is it just a means of producing children. Though procreation is a fundamental good, and, as we shall see, morally right marital intercourse must be open to procreation, the bodily unity achieved in the marital act is not a mere means in relation to it. And, most importantly here, the marital sexual act is not a just means of signifying a spiritual union distinct from it.[10] Although John Paul describes marital intercourse as a type of "language," the marital act is a *constituent* of the marriage, the marriage being a multi-leveled union *including* the bodily as well as the emotional, intellectual volitional, and so on:

> In this way the enduring and ever new language of the body is not only the "substratum," but in a certain sense, it is the constitutive element of the communion of the persons. The persons – man and woman – become for each other a mutual gift. They become that gift in their masculinity and femininity, discovering the spousal significance of the body and referring it reciprocally to themselves in an irreversible manner – in a life-long dimension (1997, p. 356).

Thus, instead of saying that sexual intercourse is a sign of mutual love – lest it be construed as an *extrinsic* sign – John Paul most often will say that it *expresses* love, or total self-giving. This expression is not extrinsic to what it expresses, but is the visible and tangible embodiment of it.

But how is the sexual act within marriage an embodiment or constitutive element of the marriage? How exactly does the sexual act complete (consummate) or renew the marriage? John Paul's answer to this question is a literal reading of the following text of *Genesis*: "Therefore a man leaves his father and his mother and cleaves to his wife, and they *become one flesh*. (2:24)" The man and the woman are each by themselves, with respect to the sexual act (the type of act that disposes them to procreation), incomplete. As a lock is incomplete without a key, so the man and the woman are themselves each only part of human nature with respect to actions suitable for procreation. Thus, in sexual intercourse they become bodily, or organically complete, and thus one.[11] And this bodily complementarity is only one level of their total personal complementarity. In marriage the bodily completion and unity in the marital intercourse is only one level of their sharing of their total lives, that is, of their total personal communion:

> Even though due to the poverty of the language, in speaking here of knowledge, the Bible indicates the deepest essence of the reality of married life. . . . . In Genesis 4:1, becoming "one flesh," the man and the woman experience in a particular way the meaning of their body. In this way, together they become almost the one subject of that act and that experience, while remaining, in this unity, two really different subjects (1997, p. 79).

And reiterating this point in this next paragraph, he explains that this bodily unity is, in marriage, an expression or embodiment of the mutual giving of their whole selves:

> We must consider that each of them, man and woman, is not just a passive object, defined by his or her own body and sex, and in this way determined "by nature." On the contrary, because they are a man and a woman, each of them is "given" to the other as a unique and unrepeatable subject, as "self," as a person (1997, p. 79).

Thus, by becoming co-subjects of a single act, an act to which they are bodily, emotionally, and spiritually oriented by their nature, the man and the woman become "one flesh."[12] And this real bodily oneness *actualizes* – not just extrinsically signifies – their total marital unity.[13]

As a bodily union, this conjugal act is also a sharing by the man and the woman of their masculinity and femininity. Speaking of the sexual act as a "sacramental sign," and as a "language of the body," John Paul says:

> The structure of the sacramental sign remains essentially the same as "in the beginning." In a certain sense, it is determined by the language of the body. This is inasmuch as the man and the woman, who through marriage should become one flesh, express in this sign the reciprocal gift of masculinity and femininity as the basis of the conjugal union of the persons (1997, p. 356).

This sharing of masculinity and femininity essentially includes a sharing of the potential fatherhood or motherhood. That is, becoming one flesh is an element of their mutual self-gift; but in becoming one flesh they dispose themselves to procreate with each other. Hence this bodily union points ahead, as it were, toward its fulfillment in the bearing and raising of children together.[14] John Paul comments on the *Genesis* text (4:1) which says, "Adam knew Eve his wife, and she conceived and bore . . . ." He speaks of the original conceiving and bearing of children as the threshold of man's history. He then says:

> On this threshold man, as male and female, stands with the awareness of the generative meaning of his own body. Masculinity conceals within it the meaning of fatherhood, and femininity that of motherhood (1997, p. 85).

And returning to this same point much later, he says:

> In this truth of the sign, and, later, in the morality of matrimonial conduct, the procreative significance of the body is inserted with a view to the future – that is, paternity and maternity (1997, p. 363).

In *Familiaris Consortio* he succinctly teaches: "Fecundity is the fruit and the sign of conjugal love, the living testimony of the full reciprocal self-giving of the spouses" (1981, no. 28).

Thus, morally right sexual acts within marriage embody marital communion, a communion which is oriented to procreation.[15] So, John Paul's teaching is distinct from the position suggested by Gareth Matthews, in his book *Body in Context.*[16] Matthews argues that one can philosophically establish only that sexual acts are *gestures* that naturally symbolize affection or care, but not necessarily marriage in the traditional sense (1992, pp. 92-109). According to Matthews, sexual acts are in the same category as embraces: they do not literally say or assert anything, but they do have a natural meaning or natural symbolism. For John Paul, however, the sexual act is much more than a gesture (though it *is* that as well). A morally right sexual act does not make present in an indeterminate way just some-union-or-other. Rather, becoming one flesh is the sort of act designed specifically to make present *marital* communion, that is, the sort of union oriented to, and fulfilled by, procreation: *that* is the kind of communion it can embody (and if it does not, then it, and the persons involved, are being *used* for extrinsic purposes). Sexual intercourse is a real, biological unity, and if it is loving and respectful sexual intercourse within marriage, it is the substratum or constitutive element of marriage: a joint act, not just a gesture, in which the two become co-subjects and thus become one. Thus, it is fundamentally a real act, and a real unification, and *because of that,* it is a gesture with profound significance or meaning.[17] So, when John Paul speaks of sex as a "language of the body;" he is not regarding the sexual act as a mere extrinsic sign. Nor is he saying only that the sexual act is a *gesture* or a natural bodily *symbol.*

6. EITHER EXPRESSION OF MARITAL COMMUNION OR IMMORAL OBJECTIFICATION

So, what is distinctive of sex according to John Paul II? What makes it so different from other bodily acts? His answer is that sexual acts are distinctive because they have the potential to embody or renew marriage, marriage being the union of a man and a woman that is characterized by its orientation to, and is naturally fulfilled by, procreation. As embodying marriage, these sexual acts have a dignity and even a sacredness. God designed human nature as masculine and feminine, and designed their differentiation as oriented to the multi-leveled, complementary union that is marriage, a union naturally fulfilled in procreation.[18] So, first, the personal communion of the masculine and the feminine, precisely as masculine and feminine, is an intrinsic and irreducible good or value. Second, this communion of the masculine and the feminine as such cannot occur except as including an openness to, and as being the sort of communion that is naturally fulfilled by, procreation. This type of union is, of course, marriage. A sexual act realizes or participates in a fundamental human good when it consummates or renews, by embodying, a marriage.[19]

And why must the sexual act be confined to marriage? Why does the sexual act involve objectification, treatment of a person as a means of satisfaction, if it is not an expression of marital communion? Extramarital sex is wrong, and involves objectification, not because such acts signify a false belief or a false proposition. Rather, they are wrong because they fail to realize any fundamental human good. Thus, what is being done cannot be the joint realization of a common good actualizing (or renewing) a communion. The act, then, can only be the pursuit of an illusory experience – pleasure or an illusory experience of union, affirmation, power, or so on – a satisfaction in one's consciousness without the realization of an actual, fundamental (and common) good.[20] It follows that such an act involves objectification and use of the other, and of one's own body, as a means to obtain that satisfaction. This understanding of the argument explains (while the misinterpretation discussed above fails to explain) why, according to John Paul, only sexual intercourse that expresses marital communion respects the unity of the body and soul (see Section 4).

In sum, the interpretation I propose can be indicated by the following text which summarizes John Paul's general argument for the basic principle in sexual ethics:

> Consequently, sexuality, by means of which man and woman give themselves to one another through the acts which are proper and exclusive to spouses, is by no means something purely biological, but concerns the innermost being of the human person as such. It is realized in a truly human way only if it is an integral part of the love by which a man and a woman commit themselves totally to one another until death. The total physical self-giving would be a lie if it were not the sign and fruit of a total personal self-giving, in which the whole person, including the temporal dimension, is present: if the person were to withhold something or reserve the possibility of deciding otherwise in the future, by this very fact he or she would not be giving totally (1981, no. 11).

Here John Paul does call the sexual act a "sign," and does say that immoral sexual acts are lies. But in the immediately preceding sentence he says that sexuality is realized in a truly human way, "*only if it is an integral part of the love by which a man and a women commit themselves totally to one another until death.*" Were extramarital sex wrong simply because it is a lie, in the sense of the signification of something one thinks or knows is not true, then there would be no point to that sentence. In other words, sexual acts *are* signs, and can be spoken of as language, but they are signs and language in a sense unique to themselves. More fundamentally, sexual intercourse is a real bodily unity, and this bodily unity is either used for purposes of private satisfaction (in acts that do not express marital communion) or is a *constitutive element or substratum, a part,* of marriage (in acts of loving and respectful marital intercourse).

So, too, the implications for specific issues regarding reproductive technologies are quite clear. John Paul's approach emphasizes the unity of the body and the soul, and thereby also the importance of the family as the appropriate environment for the bearing and raising of children. The communion of the spouses is naturally oriented to developing into family: "Conjugal communion constitutes the foundation on which is built the broader communion of the family, of parents and children, of brothers and sisters with each other, of relatives and other members of the household" (1981, no. 21). And this conjugal-familial communion is quite properly rooted in the bodily complementarity and union of the spouses: "This communion is rooted in the natural bonds of flesh and blood, and grows to its specifically human perfection with the establishment and maturing of the still deeper and richer bonds of the spirit…"(1981, no. 21).

Thus, the question whether a specific type of reproductive technology is morally right must be based on whether it is an act that is respectful of the bodily and spiritual unity of the three components of the family: sex, marriage, and children. Reproductive technologies that separate one or more from their integral unity are contrary to the fundamental good of marriage and family, and thus the fundamental good of human life (in its transmission or education). Children who come to be within a loving family come to be with bodily and spiritual connections or communions that form for them as it were a protective cocoon. Deliberately to sever the tight unity among sex, marriage, and family is to violate the respect and reverence due all of these fundamental goods (since they form, both bodily and spiritually such a unity). So, for example, *in vitro* fertilization separates procreation from the marital act, and thus is an eroding of the respect due both the child and the nature of family. The child comes to be as viewed in his or her first instant of existence as a product rather than as a person. As another example, *cloning* human beings would go even further in separating procreation from the context of sex and marriage. Cloning human beings would separate the procreation of a child not only from the marital act (which *in vitro* fertilization already does) but would go further and separate all three (sex, marriage, procreation) from each other. Contraception (immorally) severs the sexual act from any connection (such as openness) to procreation. Cloning would separate procreation not only from any particular sexual act (as does *in vitro* fertilization) but would also separate procreation from any sexual differentiation (it would be *a*sexual reproduction) and thus also make procreation in itself separate from marriage.

Franciscan University
Steubenville, Ohio

NOTES

1 Catholic teaching, however, is quite clear that there is a resurrection of the body, and thus heaven, or the completed kingdom of God, will include *both,* the spiritual and the bodily.
2 Two articles that have raised this issue are Goldman, 1977, and 1984.
3 Moreover, if Mappes were correct, then it would be logically or metaphysically impossible for someone to treat himself, or his own body, as a mere thing.

4 See Dworkin, 1974. I think the same point, however, goes for the rarer instances of pornography displaying men.

5 For an early criticism by a proponent of traditional sexual morality, see Grisez, 1964, esp. pp. 19-32.

6 In *Veritatis Splendor,* Pope John Paul states that although the teaching authority of the Church has the duty to teach that "some trends in theological thinking and certain philosophical affirmations are incompatible with revealed truth," still: "Certainly the Church's Magisterium does not intend to impose upon the faithful any particular theological system, still less a philosophical one" (1993, no. 29). And in the encyclical letter *Fides et Ratio*, John Paul teaches: "The Church has no philosophy of her own nor does she canonize any one particular philosophy in preference to others(note omitted). The underlying reason for this reluctance is that, even when it engages theology, philosophy must remain faithful to its own principles and methods. " (1998, no. 49) This reluctance to appeal to a specific philosophical system where unnecessary should be remembered when John Paul's texts are interpreted. His texts are of a different genre than a philosophical article appearing in a philosophical journal. It seems to me that this reluctance leads him often to refrain from tracing back to its roots the possible *philosophical* justification for the premises of his arguments – arguments that are given not so much to convince people not yet convinced but to *explain* or *make intelligible* the position which is held because it is part of the constant teaching of the Church. In other words, it makes it inappropriate for him to enter, at least in too much detail, the philosophical dialectical fray.

7 See also John Paul II, 1997, pp. 50, and 113.

8 The position that marriage is consummated by sexual intercourse is also held by most political communities in the west as well.

9 John Paul further writes, "The coming into being of marriage is distinguished from its consummation, to the extent that without this consummation the marriage is not yet constituted in its full reality. . . . . Indeed the very words "I take you as my wife – my husband" refer not only to a determinate reality, but they can be fulfilled only by means of conjugal intercourse" (1997, p. 355).

10 See also George and Bradley, 1995.

11 An organic action is one in which several bodily parts–tissues, cells, and so on–participate. Digestion, for example, involves several smaller, chemical actions of individual cells. But the several components of digestion form a unitary, single action. The subject of this action is the organism. So, the organism is a composite, made up of billions of parts. Its unity is manifested and understood in its actions. For most actions, such as sensation, digestion, walking, and so on, individual male or female organisms are complete units. The male or female animal organism uses various materials as energy or instruments to perform its actions, but there is no internal orientation of its bodily parts to any larger whole of which it is a part, with respect to those actions. (And this is why we recognize individual male and female organisms as distinct, complete organisms, in most contexts.) However, with respect to one function the male and the female are *not* complete, and that function, of course, is reproduction. In reproductive activity the bodily parts of the male and the bodily parts of the female participate in a single action, coitus, which is oriented to reproduction (though not every act of coitus is reproductive), so that the subject of the action is the male and the female as a unit. Coitus is a unitary action in which the male and the female become literally one organism.
One might object that the action by which the male and the female become organically one is really only completed by actual conception, and so if they do not reproduce they do not literally become one. However, if conception *does* occur that may be hours later, and whether they *now* become one cannot depend on events that occur only later. Moreover, the conditions for a successful conception are not all within the scope of their behavior. Whether a particular act of coitus results in conception depends on conditions extrinsic to the act of coitus itself. But whether their action is the sort of action that makes them one cannot depend on something wholly extrinsic to that action. And so one cannot say that the male and the female become one organism only in those acts of coitus that actually result in conception. In coitus itself – whatever may happen *after* coitus – the male and the female become one organism. Their reproductive organs are actualized, as internally designed, to be a (now) unitary subject of a single act. On this point see: Grisez, 1993, pp. 634-636.

12 Of course, not every instance of two entities sharing in an action is an instance of two entities becoming one organism. In this case, however, the potentiality for a specific type of act,

reproduction, can be actualized only in cooperation with the opposite sex of the species. The reproductive bodily parts are internally oriented toward actuation together with the bodily parts of the opposite sex. So, the same type of unity which shows that the various bodily parts of a single horse or human constitute a single organism, is found in the bodily parts of the male and the female engaging in a reproductive-type act. So they are literally, not merely metaphorically, one organism.

13 A common objection to this position is that the tradition recognizes that the sexual act of a married couple who are infertile (for example, because of age or the woman is already pregnant) is morally right. Yet such an act (the argument continues) is no different in its relation to procreation than is the sexual act of a couple (whether hetero- or homo-sexual) that is not reproductive in type. (See, for example, Matthews, 1992, pp. 161-162.) However, in the case of the infertile married couple it is only a condition extrinsic to the act that renders it infertile, and so they do become biologically one (and personally one if the act is done with respect and love). In the other case the act itself lacks what is necessary for realizing a biological union.

14 So, the fundamental goods of marriage (personal communion and procreation) should not be viewed as two distinct goods, but as two aspects or elements of the one fundamental good of marriage. On this point see: Germain Grisez, 1993, pp. 554-569; and Finnis, 1995. John Finnis shows that this position (marriage is a single basic good, communion and procreation as aspects of it) is found in St. Thomas Aquinas himself, though it seems to have dropped out in later interpretation of him: Finnis, 1998, 143-154.

15 Marriage (as is also true of marital intercourse) is not merely instrumental to procreation but is good in itself. Yet it is intrinsically oriented to, and naturally fulfilled by, procreation. Hence a marriage that does not result in children (for example one or both of the spouses are infertile) is still a genuine marriage, and is still intrinsically good, though it lacks its intrinsic natural fulfillment. On this point, see: *Vatican Council II, Gaudium et Spes,* no. 47-51; Germain Grisez, 1993, pp. 554-569; Finnis, 1997.

16 Gareth Matthews interprets John Paul's position as saying that sexual acts are speech in the strict sense, and that immoral sexual acts are wrong because they are lies in the strict sense – that is, asserting something one thinks is false with the intent to deceive (1992, pp. 94-107). I think Matthews successfully shows that these positions are mistaken (for one thing, some instances of casual sex, and most instances of commercial sex, may not involve deception). But I have argued above that to interpret John Paul in this way is inaccurate. For more on why this interpretation is inaccurate, see below, pp. XXXX.

17 Were it just a gesture with the meaning, close personal communion, then it is difficult to see why it would not be an appropriate act to share with one's parents, children, students, teachers, and so on.

18 This, of course, is not to say that marriage is merely instrumental to procreation.. Both marriage, and the act that particularly embodies it (chaste marital intercourse), are intrinsically good even if they do not reach their natural fulfillment in procreation. This point is explained well by Grisez, 1993, pp. 554-569.

19 On the notion of fundamental human goods, see *Veritatis Splendor,* nos. 50-51.

20 This point is expressed well by John Finnis, in an exchange with Martha Nussbaum, Finnis, 2002.

REFERENCES

Anderson, S. A. (2002). 'Prostitution and Sexual Autonomy: Making Sense of the Prohibition of Prostitution.' *Ethics* 112, 748-780.

Dworkin, A. (1974). *Woman Hating*. New York: E.P. Dutton.

Finnis, J. (1995). 'Law, Morality, and "Sexual Orientation."' *Notre Dame Journal of Law, Ethics and Public Policy* 9, 11-39.

Finnis, J. (1997). 'The Good of Marriage and the Morality of Sexual Relation: Some Philosophical and Historical Observations.' *American Journal of Jurisprudence* 42, 97-134.

Finnis, J. (1998). *Aquinas: Moral, Political, and Legal Theory*. New York: Oxford University Press.

Finnis, J. (2002). 'Is Homosexual Conduct Wrong? A Philosophical Exchange,' in A. Soble (Ed.), *The Philosophy of Sex, Contemporary Readings , 4th edition*. Lanham, Md.: Rowman and Littlefield. 97-102.

George, R.P. and G.V. Bradley (1995). 'Marriage and the Liberal Imagination,' *Georgetown Law Review* 84, 301-320.

Goldman, A. (1984). 'Better Sex,' in R. Baker and F. Elliston (Eds.), *Philosophy and Sex, 2nd ed.*. New York: Prometheus.

Goldman, A. (2002). 'Plain Sex,' in A. Soble (Ed.), *The Philosophy of Sex, Contemporary Readings , 4th edition*. Lanham, Md.: Rowman and Littlefield. 39-55.

Grisez, G. (1964). *Contraception and the Natural Law*. Milwaukee: Bruce Publishers.

Grisez, G. (1993). *The Way of the Lord Jesus, Volume 2, Living a Christian Life*. Quincy, IL.: Franciscan Press.

John Paul II (1981) *Familiaris Consortio*. (The Christian Family in the Modern World). Available at: www.vatican.va/holy_father/john_paul_ii/apost_exhortations/documents/hf_jp-ii_exh_19811122_familiaris-consortio_en.html.

John Paul II (1993). *Encyclical Letter Veritatis Splendor* (The Splendor of Truth). Available at www.vatican.va/holy_father/john_paul_ii/encyclicals/documents/hf_jp-ii_enc_06081993_veritatis-splendor_en.html.

John Paul II (1997). *Theology of the Body*. J.S. Grabowski (Ed.), Boston: Daughters of St. Paul.

John Paul II (1998). *Fides et Ratio* (Faith and Reason). Available at http://www.vatican.va/holy_father/john_paul_ii/encyclicals/documents/hf_jp-ii_enc_15101998_fides-et-ratio_en.html.

Mappes, T. (2002). 'Sexual Morality and the Concept of Using Another Person.' in A. Soble (Ed.), *The Philosophy of Sex, Contemporary Readings , 4th edition*. Lanham, Md.: Rowman and Littlefield. 207-224.

Matthews, G., OP. (1992). *The Body in Context, Sex and Catholicism*. London: SCM Press.

Nussbaum, M. (2002). 'Objectification,' in A. Soble (Ed.), *The Philosophy of Sex, Contemporary Readings , 4th edition*. Lanham, Md.: Rowman and Littlefield.

Rahner, K. (1986). *Foundation of Christian Faith, An Introduction to the Idea Of Christianity*. W.V. Dych (Trans.), New York: Crossroad.

Vatican Council II (1965). *Gaudium et Spes* (*The Constitution of the Church in the Modern World*). Available at http://www.vatican.va/archive/hist_councils/ii_vatican_council/documents/vat-ii_cons_19651207_gaudium-et-spes_en.html

CHAPTER EIGHT

MICHAEL J. MURRAY

PROTESTANTS, NATURAL LAW, AND REPRODUCTIVE ETHICS

There is little doubt that the life and work of John Paul II has been more intensely watched and appreciated by Protestant thinkers than that of any other Pope since the Reformation. One would hardly have expected this to be the case back in the fall of 1978 when Cardinal Karol Jozef Wojtyla was elevated the chair of Saint Peter. With the influence of the Roman Catholic Church apparently on the wane in the West, none could have anticipated that John Paul II would become not only one of the most important world leaders of the twentieth century, but also one of the leading theological, philosophical, political, and social theorists of our time. His teaching in these areas has deserves sustained and careful reflection by all Christians.

In this essay I would like to offer some distinctively Protestant reflections on the essays by Laura Garcia and Patrick Lee in which they aim to tie together some of the important teaching of John Paul II on the topics of marriage, sexuality, and reproduction. I begin with some general remarks on the stance of Protestants on the natural law tradition that underlies John Paul II's teaching and, more broadly, Catholic moral theology. I then consider some of the arguments offered by Garcia and Lee that might be less readily received by Protestants.

1. PROTESTANTS AND NATURAL LAW

Natural law theories have been developed in ways that involve quite diverse metaphysical and epistemological convictions. Nonetheless, they share in common the claim that there are certain states of human nature which suffice to define moral norms for human action. For theists, the natural law consists in those moral norms of conduct which God has prescribed to us by means of the nature he created in us. But we can begin to unpack the theory without making any direct reference to God or creation.

Most of us would readily acknowledge that the practices of eating-and-purging found among bulimics is wrong. Eating and voluntarily purging is not merely unhealthy, it is a flagrant degradation of bodily integrity. We know this because we know enough about the human body and about human physiology to know that behaving in this way seriously disrupts the proper functioning of the body. Bodies are constructed in such a way that we eat to secure nutrition and hydration. Voluntarily purging with the aim of denying the body nutrition and hydration thus

C. Tollefsen (Ed.), John Paul II's Contribution to Catholic Bioethics, pp. 121–129.

serves to thwart the proper functioning of our bodies. It is for this reason, perhaps among others, that we would all regard eating and purging as bad.

But without emendation, it is hard to see how the badness of the action amounts to much more than pragmatic irrationality. We all are bound by principles of pragmatic reason to pursue what is good for us and to avoid what is bad. Since eating and purging thwarts the proper function of our body, it is bad for us and thereby pragmatically irrational.

Can we take this opening move further so that facts about human nature and proper functioning can serve not only to generate pragmatic norms, but also to generate genuinely *moral* norms? Perhaps the naturalist advocate of natural law theory can argue that this sort of pragmatic irrationality is, on its own, sufficient to constitute immorality.[1] But it is not obvious how such an argument would go. Undoubtedly it is harmful to be stupid. But stupidity and wickedness seem, on the surface at least, to be different commodities.

The theist, however, can say more. Since God is the creator of all that there is, things are what they are, and do what they do, because of the nature God has given them. On this view, proper functioning is not merely a natural fact about created entities, but a supernaturally ordained fact. As a result, to act so as to thwart proper function is to act so as undermine the integrity of the divine creation. Thus, when we act in such a way that we thwart the proper functioning of the creation, whether by engaging in activities that thwart our own proper functioning, or by engaging in activities that thwart the proper functioning of other parts of creation, we not only act in ways that are pragmatically irrational, but we also offend the creator. From this, the theist can argue that running afoul of the norms of proper function established by our natures is not only pragmatically irrational, but also *sinful*.[2]

Furthermore, since the moral norms in view here are derived from facts about human nature they are discoverable by human reason insofar as human beings are capable of understanding their own nature and the nature of human proper function. Thus, insofar as moral norms are grounded in our natures, all human beings, even those who lack access to special divine revelation, are at least in principle capable of knowing a wide range of moral truths.

Nothing about natural law theory commits one to saying that creaturely natures suffice to ground the truth of all moral norms. In addition, there might be, indeed all Christians think there are, moral norms that arise from positive divine law (that is, where God commands creatures to do things which are not strictly required by their natures). It is not important here to spell out all of the sources of moral normativity that might be endorsed by the Christian. It is only important to see that the natural law theorist need not be committed to the claim that created natures are the only source of such normativity.

Protestants have traditionally objected to natural law theories, sometimes on metaphysical grounds, sometimes on epistemological grounds, and sometimes on Biblical grounds. On the metaphysical side, two very different sorts of critiques have been leveled. The first critique is offered by those who object to the very idea of an objective human nature, created by God in every human creature. These anti-essentialists have no patience with natural law since, by their lights, the needed foundation is missing. But even those with more traditional metaphysical leanings might be troubled here. Many Reformed theologians are concerned that natural law theory removes God as the grounds for moral norms, and replaces it with a human nature that is fallen to such an extent that either it cannot provide adequate resources for coming to know those norms or, even worse, is no longer capable of providing the proper metaphysical grounding for them.

Responding to the anti-essentialists is too great a task to attempt here. To the second criticism, two things might be said. First, the claim that natural law theory inappropriately removes God from the grounds of moral normativity is too quick.[3] Most of those who raise this criticism fail to see that even on natural law theory, the norms that arise from consideration of human nature and proper function have moral force because they are chosen and instantiated by God as devices for defining the nature of goodness and proper function in the created order. In this way, God's creative intentions ground or underpin, albeit indirectly, the truth of moral claims in a way analogous to the way in which my intentions explain or ground the fact that it is wrong for you to use my candy dish as a spittoon, or my log house for firewood. I designed and created these things for certain purposes, and failing to respect those purposes is sufficient to constitute your action as wrong.

But this solves only half of the problem. The Protestants I describe above argue not only that natural law theory removes God as the proper grounds of moral normativity, but also that it replaces God with a human nature that is so utterly corrupted by sin, that it is no longer capable of acting as truthmaker for the norms.[4] A number of things might be said in response to this objection. First, one might agree that the *imago dei* is so utterly corrupted in human beings that it no longer can ground genuine moral norms, but go on to claim that what then grounds moral norms are facts about what uncorrupted human nature was or would be like. Unfortunately, one who wishes to adopt this line will have a correspondingly harder time defending the claim that natural law theory provides us with resources for discovering the moral truth without special divine revelation. How can we know, the critic might ask, what the uncorrupted nature would look like absent revelation?

I think the more plausible response is simply denial. Surely we think that *some* moral norms can be discerned from facts about human nature and proper function. Eating-and-purging and rape provide personal and interpersonal examples. Furthermore, most of the Protestants conservative enough to press this line of reasoning would also be likely to agree that St. Paul grounds the wrongness of homosexual activity in the fact that it is contrary to nature ("men abandoned *the natural function* of the woman and burned in their desire toward one another, men with men committing indecent acts and receiving in their own persons the due penalty of their error" (Romans 1:49-50).[5]

On the epistemological side, once again, two main objections have been leveled. First, some have argued that natural law theory diminishes the importance of or need for special revelation (Henry, 1995, p. 56). Others have pressed an argument closely related to the metaphysical objection discussed above. They argue that even if human fallenness does not entail that human nature is utterly corrupted, it does entail that human wickedness renders moral truth utterly opaque to us. It is thus an illusion to think that we could ever derive moral truths from consideration of the human nature, however corrupt or uncorrupt it might be (Henry, 1995, p. 59).

Both of these concerns can be deflected. It is true that many advocates of natural law believe that we can largely discover the moral truth by unaided reflection on human nature, but this does not suffice to undermine the importance of special revelation. The first reason for this, as we have seen above, is that there is reason to think that the moral truth is not exhausted by consideration of those norms that derive from human nature. The Israelites were morally obliged to offer the sacrifices prescribed in the Mosaic Law. But these moral obligations sprang not from human nature, but from revealed, positive divine law. Second, even when we consider those moral norms which are grounded in human nature, revelation might play a crucial role in keeping us from error. Undoubtedly, the history of ethical reflection on human nature and the norms it does or does not support has been vexed. Appeals to human nature have been used to ground the ethical permissibility of oppression, genocide, slavery, perversion, and a host of moral atrocities. Thus, as with natural theology, it is perfectly reasonable to think that part of the role of special revelation is to counteract the human tendency towards error in these matters.

Finally, this position seems well supported by Scripture when St. Paul writes, for example, that the Gentiles who have no access to special divine revelation, have a law of conscience written on their hearts, "For when Gentiles who do not have the Law do instinctively the things of the Law, these, not having the Law, are a law to themselves, in that they show the work of the Law written in their hearts, their conscience bearing witness and their thoughts alternately accusing or else defending them" (Romans 2:14-15).

In addition, it is reasonable to think that the nature of goodness more generally is available to unaided human reason since Paul, in the same letter, explains that the divine nature itself can be grasped by reflection on the created order: "For since the creation of the world His invisible attributes, His eternal power and divine nature, have been clearly seen, being understood through what has been made, so that they are without excuse" (Romans 1:18-20).

Nonetheless, the Protestant critic's emphasis on human fallenness, even when short of affirming total depravity, raises important cautions about natural law. Natural law theorists differ over the question of how we are to discover the natural law. Some, whom we might call "derivationists," adopt a top down approach, arguing first for a conception of human nature and the human *telos*, and then defining proper function and well-being in terms of these. Others, whom we might call "intuitionists," argue, in bottom up fashion, that the way we come to grasp the fundamental human goods is by considering those things to which human persistently direct their actions, perhaps across times and cultures. Such persistent directedness indicates that the objects of pursuit are indeed those goods at which human action aims. So, if we notice that, across times and cultures, human beings tend to pursue love, friendship, social harmony, etc., we can infer that these are among the activities that constitute human flourishing and proper function.[6] More plausible, perhaps, is a mixed approach. It is hard to see how one might discern the truth about human nature without paying some attention to the forms of activity that human beings persistently pursue. And it is likewise difficult to see how we might be able to affirm the goodness of certain persistent desires without some independent conception of what things serve to promote or harm human flourishing.

Insofar as one adopts intuitionism, the epistemological Protestant critique at least serves to raise some concerns. It is hard to read some of the stern New Testament warnings about the depth of human moral corruption without becoming skeptical about our ability to grasp the truth about human nature and human flourishing unaided, especially if we derive them from persistent human desire. Consider, for example, Paul's remarks in Romans:

> There is no one righteous, not even one; there is no one who understands, no one who seeks God. All have turned away; they have together become worthless; there is no one who does good, not even one. Their throats are open graves; their tongues practice deceit. The poison of vipers is on their lips. Their mouths are full of cursing and bitterness. Their feet are swift to shed blood; ruin and misery mark their ways, and the way of peace they do not know. There is no fear of God before their eyes. (Romans 3:10-18)

Paul's description of the tendencies of the human heart should at least instill some doubts about the extent to which human desires are capable of tracking genuine goods – the sorts of goods which serve to promote genuine flourishing.

Similar cautions are in order concerning derivationism. Just how much caution is in order depends on how radically one takes human nature to be deformed by the fall. For the most extreme Calvinists, the prospects are grim. And such Calvinists are likely to argue that their worries are vindicated by recent discussions of homosexuality, gay marriage, sexual ethics, and so on where popular moralists have proclaimed every form of sexual activity under the sun to be both "natural" and good.

Yet in light of the considerations noted earlier, it seems that the prudent path is to assume that the deliverances of our reflection on natural law will be erroneous or at least incomplete absent supplementation and correction by revelation.

In sum, stiff Protestant resistance to natural law theories seems ultimately misplaced. Protestants have been more emphatic about the role of the fall in both distorting human nature and disabling human capacities for unaided reflection on ethical matters. But these concerns ultimately serve to show only that Protestants will be unwilling to admit that considerations of natural law can, on their own, provide one with the full content and complete confidence concerning the moral truth.

2. GARCIA AND LEE ON MARRIAGE AND SEXUAL ETHICS

In their essays, Laura Garcia and Patrick Lee focus their attention on the works of John Paul II, both preceding and during his papacy, which bear on the moral dimensions of marriage, sexuality and reproduction. These reflections are refreshing in an era in which marriage is reduced, at best, to another mere contractual relationship among consenting individuals. On the view they develop, the marital union is one in which both husband and wife commit themselves to a genuine and selfless love which seeks both to secure their union and to promote the well-being of the other as a person.

Both argue that this conception of marriage has significant implications for sexual and reproductive morality. On the one hand, it weighs against the moral permissibility of sexual activity outside of marriage, and against treating one's spouse merely as an object to be used for one's own pleasure. Both of these activities intrinsically violate the loving commitment intrinsic to marriage – the first since it gives to another that which is pledged to one's spouse alone, the other because it fails to treat one's spouse as an end and thus as a proper object of love.

More controversially, Garcia and Lee extend this conception to support arguments against contraception and the use of artificial reproductive technologies. Concerning the former, Garcia remarks,

> Genuine love for one's spouse requires openness to procreation in sexual relations. Human persons are sexual beings, and the body and its powers are part of their glory. To love a human person, to truly desire his or her good, requires an acceptance of the body as well – its natural gifts and capacities. . . . The norm of love does not require that couples intend in each sexual act to conceive a child, but that this intrinsic power proper to the other person in his or her body is not deliberately opposed or negated (Garcia, p. 98).

Similarly, Lee argues,

> As embodying marriage . . . sexual acts have a dignity and even a sacredness. God designed human nature as masculine and feminine, and designed their differentiation as oriented to the multi-leveled, complementary union that is marriage, a union naturally fulfilled in procreation. So, first, the personal communion of the masculine and the feminine, precisely as masculine and feminine, is an intrinsic and irreducible good or value. Second, this communion of the masculine and the feminine as such cannot occur except as including an openness to, and as being the sort of communion that is naturally fulfilled by, procreation. This type of union is, of course, marriage. A sexual act realizes or participates in a fundamental human good when it consummates or renews, by embodying, a marriage (Lee, p. 115).

The position defended here represents the standard, traditional view as developed by the Roman Catholic church, especially since Pope Paul VI's 1968 encyclical *Humana Vitae*. There Pope Paul VI argues quite plausibly that sexual union in marriage has both a unitive and a procreative function. Understood this way, one might think that just as eating and purging thwarts human proper function and is thereby wrong, so engaging in sexual activity with one's spouse is wrong when it thwarts either the unitive or procreative functions of the conjugal act.[7]

But this is to argue too quickly. It is true that an argument of this sort might show that we have a *prima facie* obligation to seek to promote both unitive and procreative functions of the conjugal act. However, to make this argument work, it must be the case that actions which seek to thwart one or the other of these functions are intrinsically disordered. It is hard to see how such an argument might go.[8] To see the difficulty, we can once again appeal to the analogy of the bulimic. It is true that eating and voluntarily purging is a *prima facie* wrong. But there are cases where doing so seems rather to enhance the proper functioning of the whole person, e.g., when the person has recently ingested a toxin. Likewise, in cases where a couple has more children than they can support, or where the woman's life or physical well-being is severely threatened by future pregnancy, or where the risk of serious illness for a future offspring is overwhelmingly high, etc., the *prima facie* good of the procreative function of the conjugal act seems *ultima facie* overridden.

Nevertheless, even if the Protestant might think that the natural law theorist has overreached a bit in the argument, it is worth noting that there is a line of argument here that has not been sufficiently appreciated by Protestants. While it may be true that an outright ban on contraception is more than is warranted, flagrant disregard for the procreative function of the conjugal act is also unwarranted. Protestants have been liable to underappreciate the force of this argument. Recall that the underlying issue here is whether or not we can make sense of the idea of the claim that God creates human beings with a nature which specifies, at least in part, the character of human proper function and well-being. The answer to that question seems to be yes. And insofar as that is right, all Christians are obliged to admit the *prima facie* goodness of promoting that proper function. It seems inescapable that procreation is at least one of the outcomes at which conjugal union naturally aims. As a result, unless there is overriding reason for not doing so, Protestants should likewise agree that contraception is immoral. And the standard for what counts as an overriding reason is not trivial. This is important since so often contraception is used in order to serve greedy and selfish desires, rather than to secure the genuine, overriding well-being of the couple or family.

Garcia and Lee further make the case against the use of artificial reproductive technologies (ART's). It seems clear that some ART's are immoral, though for quite varying reasons. Some, like many surrogate pregnancy arrangements, seem intrinsically exploitive or degrading; others, like many forms of *in vitro* fertilization, involve illicitly taking the life, or other unjust treatment, of the embryo. Natural law theorists typically look to ground their objections to ART's in at least one of two considerations. Garcia describes the two as "the dignity of the child and the dignity of the spouses in the sexual union"(Garcia, p. 100). We can consider these separately, beginning with the second.

While there is no doubt that some forms of ART do undermine the dignity of the spouses in their sexual union, it is not clear that this is always the case. Some forms of ART might serve, in fact, to *enhance* proper functioning in the marital union. To pick the most extreme example, we might consider cases in which the male ejaculate produced during intercourse is harvested and used to fertilize a single egg which is then transferred to the wife's fallopian tubes. Here we have employed an artificial technology as a means of enhancing the procreative function of the marital union. Undoubtedly ART's are not typically used in this very fashion. But the point is simply that there does not seem to be anything about a procedure like this one which detracts from proper function or which undermines spousal dignity, and this invites further reflection on just what sorts of interventions serve to enhance procreative function and which detract from it.

Of course this does not take into account the other line of argument often used in this context, and referred to by Garcia. That is, one might argue that the problem here is rather that ART's fail to respect the dignity of the child. Some argue that this is so because the child has a right to come to be through the fruit of conjugal union, others argue that ART's treat the child as a commodity rather than as a person.[9] Neither of these arguments is especially convincing. It may be true that the child is entitled to come to be as the result of an act of loving commitment on the part of a husband and a wife. But why must the act be an unmodified instance of conjugal union? We might also ask in what sense the parents described in my example above can be thought of as failing to respect the personhood of the child, or as treating the child as a commodity. It is hard to see how seeking to conceive of the child in this manner commodifies the child any more than would the use of fertility drugs. In that case, the drugs serve to enhance the procreative capacities of the woman, and thus the procreative function of the conjugal act. In this way, the couple uses artificial means to enhance the odds of conceiving a child. But this is not to treat the child is a commodity, and it is hard to see how things are different in the IVF case I describe above.

3. CONCLUSION

Undoubtedly, Protestant philosophers and theologians have a great deal to learn from the rich moral teaching of John Paul II and from the Roman Catholic natural law tradition. But as the above makes clear, there is room for substantive disagreement and dispute about exactly what moral norms arise from consideration of human nature. I have argued that Protestants should be open to the notion that natural law provides a fruitful source for moral reflection, at least when guided and supplemented by special divine revelation. Insofar as our unaided reflections on natural law lead us to conclusions at odds with the declarations of Scripture, so much the worse for our unaided reflections.

Most Roman Catholics would find such sentiments agreeable. Where disagreement is more likely to arise is in those cases where the magisterium of the Church has made declarations about the deliverances of natural law which are less than obvious. In these cases, Protestants will likely feel that their Roman brethren have become beholden to a standard that they cannot endorse. It is here, however, where fruitful philosophical dialogue is likely to emerge. It is this sort of dialogue I commend.

Franklin and Marshall College
Lancaster, Pennsylvania

***

Research on this essay was partially supported by grants from the University of Notre Dame Center for Philosophy of Religion and from the American Philosophical Society.

NOTES

1. This is a somewhat unfair way of characterizing the view. Defenders of this position claim that facts about human nature not only make true certain claims about proper function, but also norms with intrinsic moral authority. This was the view of St. Thomas Aquinas, a view criticized by John Duns Scotus, among others. Among contemporaries, the view that natural law has intrinsic moral authority is defended by Finnis, 1980, and criticized by Adams, 1999.
2. This line of reasoning is by no means uncontroversial. Divine command theorists might argue that facts about proper function are insufficient to ground moral norms even if the natures that establish proper function are instantiated intentionally by God. By their lights, moral force springs only from the divine command itself. These are issues that deserve more careful treatment than I can give them here. The reader is advised to consult Foot, 2001, esp. pp. 66-80, and Murphy, 2001, esp. pp. 222-7, for further discussion of these and related issues.
3. For one who makes this claim see Henry, 1995, p.56.
4. See, for example, Leithart, 1996.
5. All citations are from the New American Standard Bible, Foundation Publication Inc.
6. The terms "intuitionist" and "derivationist" come from Murphy, 2001, Chapter One.
7. The argument against contraception is sometimes put in terms of thwarting or failing to respect the masculinity and/or femininity of the married couple. These features are intrinsic to the nature of the person and are intrinsically oriented towards fertility. I will not treat this version of the argument here, but it should be noted that the response I will give to the unitive/procreative function argument above can be made equally well against this argument *mutatis mutandis*.

8 In fact, Garcia recognizes the need for bolstering the argument in just this way. Shortly after the passage cited above she adds, "Some theologians argue that recourse to artificial means of preventing conception can sometimes be justified in cases where a married couple has a general willingness to accept children (at some time or other), and when the marital act is motivated in part at least by morally good intentions – for example, to give pleasure to each other, to express love, to console or encourage each other and so on. But this rationale assumes that a contracepted sexual act, when freely chosen, is not in itself (as an intended act) unjust or unloving. If the act is otherwise morally good or neutral, then of course it is morally justifiable in cases where no vicious intentional states enter into it; but if it is a morally objectionable act, then further motives and ends cannot serve to justify it. We can see this more clearly perhaps in the case of acts that everyone recognizes as unjust, such as date rape (coercing a person into sexual intercourse without her or his consent). Whatever 'loving' purposes may lie behind someone's choice of this act, the decision to commit *this act* is unloving. *One cannot perform an unloving act lovingly*, though one may see it as a *means* to some further morally benign *end*" (Garcia, p. 99). However, Garcia here merely assumes that the contracepted sexual act is intrinsically disordered. This is something that requires an independent defense against the line of reasoning I offer presently.

9 In fact, Garcia invokes both considerations in the following: "In the first place, [the 1987 Vatican document *Donum Vitae*] argues that an injustice is done to the human being produced in this way, for the very reason that he or she becomes a product. 'It is through the secure and recognized relationship to his own parents that the child can discover his identity and achieve his own proper human development' (*DV* II.1). Children naturally conceived are meant to be the fruit of marital love – an occasion for joy, received as a gift to their parents, a living sign of their love and union. Although the injustice is more obvious in cases where the child is not genetically tied to both parents, it is also true of the child conceived by technical processes distinct from the marriage act. A child 'cannot be desired or conceived as the product of an intervention of medical or biological techniques; that would be equivalent to reducing him to an object of scientific technology' (DV II.5) wherein his coming-to-be is evaluated by standards of technical efficiency and is under the direct control of other persons" (Garcia, p. 101).

REFERENCES

Adams, R. (1999). *Finite and Infinite Goods*. Oxford: Oxford University Press.

Finnis, J. (1980). *Natural Law and Natural Rights*. Oxford: Oxford University Press.

Foot, P. (2001). *Natural Goodness*. Oxford: Oxford University Press.

Garcia, L. (2004). 'Protecting Persons.' In *John Paul II's Contribution to Catholic Bioethics*. C. Tollefsen (Ed.), The Netherlands: Kluwer Press. 93-105.

Henry, C.F.H. (1995). 'Natural Law and a Nihilistic Culture.' *First Things*, 49 (January 1995).

Lee, P. (2004). 'The Human Body and Sexuality in the Teaching of John Paul II.' In *John Paul II's Contribution to Catholic Bioethics*. C. Tollefsen (Ed.), The Netherlands: Kluwer Press. 107-120.

Leithart, P.J. (1996). 'Natural Law: A Reformed Critique.' *Premise*, III(2).

Murphy, M. (2001). *Natural Law and Practical Rationality*. Cambridge: Cambridge University Press.

CHAPTER NINE

ANDREW LUSTIG

JOHN PAUL II ON THE GOOD OF LIFE

1. INTRODUCTION

According to one commentator, the fundamental theme that hallmarks the thought and writings of John Paul II is "the indwelling of God in each human being and consequently in human society" (Donders, 1996, p. ix). The sequential ordering of that indwelling is decidedly important. John Paul's philosophical approach is characteristically phenomenological and personalist, as a matter of his own history and educational influences. Given his experience as a Polish citizen and churchman, John Paul was well acquainted with the dangers of communism as a collectivist ideology that usurped the central functions of individual freedom and responsibility. John Paul's advanced education combined Neo-Scholastic methods in moral theology with the philosophical phenomenology associated primarily with Max Scheler. From these influences, he developed a robust ethical vision with the concrete reality of persons at the heart of all deliberations.[1] While John Paul, then, reiterates a characteristically Catholic emphasis on persons as *necessarily* in community (indeed, sociality is a constitutive dimension of personhood), his own discussion has stressed the concrete requirements of human dignity. Indeed, in his very first encyclical, *Redemptor Hominis*, John Paul enunciates the foundational commitment of his pastoral and theological worldview: "that the human being is the primary and fundamental way for the Church" (quoted in Curran, 2002, p. 62). In his first social encyclical, *Laborem Exercens*, his emphasis on the priority of labor over capital is best viewed as the practical corollary of his personalist approach: labor, unlike capital, is not merely instrumental, but a reality that expresses the status of the laborer as a free subject, as one made in the image and likeness of God, and therefore as one who bears "an incomparable dignity and the rights that flow from that dignity" (O'Brien and Shannon, 1992, p. 447).

In considering John Paul's overall corpus, one is struck especially by the way that his account of personal freedom is linked to the demands of justice, as expressed through the language of individual rights. From his earliest writing on the dignity of labor, through his well-known encyclical *Centesimus Annus* on the occasion of the hundredth anniversary of *Rerum Novarum*, and in his broader, culturally critical encyclicals, especially *Veritatis Splendor* and *Evangelium Vitae*, John Paul has reaffirmed the central development in social teaching since the time of John XXIII: viz.,

C. Tollefsen (Ed.), John Paul II's Contribution to Catholic Bioethics, pp. 131–150.

that individual rights, both civil and economic, are the necessary concomitants of human dignity.

While he has seldom addressed in detail the full range of entitlements that flow from theologically grounded claims of dignity, his fundamental theological and philosophical convictions have implications for a range of issues in bioethics. In that light, I will therefore structure this essay in four sections. First, drawing primarily on John Paul's discussion of certain bioethics issues in *Veritatis Splendor* and *Evangelium Vitae*, I will set out six core themes that John Paul develops and suggest that each theme has implications for matters of general ethical method and for specific bioethics topics.

I will then, in my second and third sections, turn directly to two particular issues that illuminate key features of John Paul's understanding of persons and their exercise of responsible freedom. In the second section, I will consider euthanasia, comparing John Paul's analysis of that issue with the central themes in a number of recent secular discussions. John Paul's rejection of assisted suicide and euthanasia as legitimate choices, while reflecting Catholicism's traditional proscription, also functions in the larger context of his prophetic and critical posture against key elements of what he sees as "culture of death." In addition, it offers a clear contrast to accounts that emphasize individual freedom unmoored from the shared moral constraints that follow from natural law reflections.

In the third section, I will consider a recent statement by the U.S. Bishops regarding the right to health care as an entitlement due individuals in light of their dignity as creatures made in the image of God. Here my discussion will extrapolate from the general features of John Paul's discussion to more focused analysis provided by the bishops. By focusing on two issues that are usually analyzed in isolation – one an issue of putatively "individual" choice, the other a right owed to individuals as a communal obligation – I intend to show the underlying unity of John Paul's approach. Claims of justice are not to be isolated from the fundamental moral obligations of individuals, and vice versa. The language of "rights" and the language of "right," then, are indissoluble in John Paul's understanding, because the choices of individuals within society and obligations incumbent on society toward individuals are both rooted in fundamental features of human existence, as captured in the traditional language of natural law.

Nonetheless, in my conclusion, I highlight several conceptual unclarities and tensions that continue to plague Catholic social teaching, including the writings of John Paul II. These unclarities raise questions about the cogency of social teaching as a whole even as they indicate an agenda for future discussion and development.

2. SIX FUNDAMENTAL THEMES

2.1 The Nature and Scope of Human Freedom

As a basic theme throughout his writings, John Paul offers an account of freedom that stands in contrast to many secular understandings. Freedom is not, for John Paul, an absolute or unfettered value; rather, freedom in its deepest sense, can only be viewed in the light of truth. That truth, as discussed in *Veritatis Splendor*, is acknowledged as the truth of Christ, indeed, the truth which is Christ. This freedom, moreover, is an embodied freedom, not some gnostic rejection of material existence. The facts of embodiment are pivotal in interpreting and asserting key natural law claims about the purposes of rightly ordered human choices and actions, and about the licitness of particular actions (especially concerning birth and death) as expressing or distorting the core unity of persons as inspirited bodies, embodied spirits.

2.2 The Connection Between Faith and Morality

Throughout his theoretical and pastoral writings, John Paul emphasizes the necessary connection between faith and morality. Faith cannot be separated from morality nor, given the Church's function as moral teacher to both the faithful and the world at large, can a properly interpreted morality be separated from faith. This theme, of course, raises important issues concerning how best to understand the Church's teaching authority, how Revelation and reason interact as sources of moral authority, and how these sources are to be interpreted in drawing moral conclusions about specific actions.

2.3 The Commitment to Objective, Universal Truth

Throughout his writings, John Paul emphasizes that the Church's teachings on moral issues are grounded in a commitment to objective, universal truth, most fully manifested in the person of Jesus Christ. This same alliance between genuine moral freedom and the reality of objective moral truth also functions in natural law epistemology in a more "minimal" sense than its full expression in Christ. Faith based on revelation and a natural human morality based on "right reason" (*i.e.*, practical reason teleologically directed toward proper human ends) are not compartmentalized in Catholic teaching in either its moral theology on particular issues or its social teaching. For example, in *Veritatis Splendor*, John Paul asserts the links between natural law as a source of moral insight and the goods that human beings share in common:

> Precisely because of this truth, the natural law involves universality. Inasmuch as it is inscribed in the rational nature of the person, it makes itself felt to all beings endowed with reason and living in history. In order to perfect himself in his specific order, the person must do good and avoid evil, be concerned for the transmission and preservation of life, refine and develop the riches of the material world, cultivate social life, seek truth, practice good, and contemplate beauty ... inasmuch as the natural law expresses the dignity of the human person and lays the foundation for his fundamental rights and duties, it is universal in its precepts and its authority extends to all mankind (1993, no. 52).

This third emphasis, of course, draws again upon the first theme, this time through an explicit acknowledgment of natural law. In this passage, John Paul underscores that freedom cannot be an end in itself, unfettered to rightly ordered choice, because as Christians we acknowledge that we participate in a teleologically shaped created order. We have ends that we can know and goods we should pursue that we must never choose directly against.

2.4 The Church's Role as Moral Teacher

In his writings, especially *Veritatis Splendor* and *Evangelium Vitae*, John Paul stresses that morality must be understood according to Revelation and reason, and in light of the Church's magisterial authority as teacher. In *Veritatis Splendor*, for example, John Paul reiterates specific conclusions about the nature of certain acts that the Church has traditionally deemed to be intrinsically evil; i.e., acts that allow for no justified exceptions. Positive precepts (for example, to worship God, to honor one's parents) "which order us to perform certain actions and cultivate certain dispositions, are universally binding; they are unchanging." So, too, the negative precepts of the natural law are "universally valid. They oblige each and every individual, always and in every circumstance." As examples of the latter, John Paul observes that "The Church has always taught that one may never choose kinds of behavior prohibited by the moral commandments expressed in negative form in the Old and New Testaments" (John Paul II, 1993, no. 52). Suicide and assisted suicide, as *mala in se*, are among such universally proscribed choices and actions.

To be sure, such statements about the unchanging and universally binding nature of the negative precepts of the natural law leave a great deal unaddressed, especially when one considers the licitness of particular issues in bioethics, including sexual ethics. To speak of the moral law as "in principle" accessible to human reason is, on Roman Catholic terms, unobjectionable. But how that moral law will be accessible to all persons everywhere at all times, independent of a Revelation-informed faith, and independent of the Church's teaching authority, is hardly obvious. Here the issue goes to the heart of ecclesiological understanding, and the way that respective sources of moral authority are drawn upon to reach moral conclusions, even about putatively shared natural law precepts. To large extent, even in John Paul's more systematic discussions, the bridgework between basic moral prohibitions that he finds in the second table of the Decalogue and specific judgments about particular acts is not provided but assumed.

2.5 The Fundamental Option and Individual Moral Decisions

As noted above, John Paul's writings are infused with a personalism that keeps intact the link between one's basic orientation, the so-called "fundamental option," and the way that this stance is expressed in particular moral decisions. As an extension of the principle of "embodiedness" and a pointed rejection of a dualism that would separate one's fundamental convictions from the way that those convictions are expressed, *Veritatis Splendor* calls for a unitary perspective on morality:

> To separate the fundamental option from concrete kinds of behavior means to contradict the substantial integrity or personal unity of the moral agent in his body and in his soul .... Judgments about morality cannot be made without taking into consideration whether or not the deliberate choice of a specific kind of behavior is in conformity with the dignity and integral vocation of the human person (1993, no. 67).

2.6 The Claims of Conscience in the Moral Life

As a sixth theme, John Paul sets out a nuanced understanding of the way that claims of conscience function in the moral life. In keeping with the first theme – that responsible freedom must be yoked to truth – John Paul consistently argues that conscience, though freely exercised, is not self-validating. Freedom of conscience must be understood in light of the object of choice toward which conscience is directed. *Veritatis Splendor* speaks directly to this point:

> The reason why a good intention is not itself sufficient, but a correct choice of actions is also needed, is that the human act depends on its object, whether the object is capable or not of being ordered by God, to the One who alone is good, and thus brings about the perfection of the person...Christian ethics, which pays particular attention to the moral object, does not refuse to consider the inner teleology of acting, inasmuch as it is directed to promoting the true good of the person; but it recognizes that it is really pursued only when the essential elements of human nature are respected (1993, no. 78).

This final theme recurs throughout John Paul's writing – the truth that the moral licitness of actions is not determined by intentions alone, that morality requires an intentionality in keeping with the appropriate object of choice, and that the true good of the person requires a respect for the essential elements of human nature. As before, however, problems of specification will emerge vis-a-vis particular questions, viz., what are the factors relevant to accurately describing the object of moral choice? A theme that is unexceptionable as a general claim about conscience becomes more controversial when applied to particular cases, depending on how the legitimate objects of choice and the essential elements of human nature are described in each instance.

3. THEOLOGICAL ARGUMENTS AGAINST VOLUNTARY ASSISTED SUICIDE AND EUTHANASIA

John Paul's most extensive discussion of euthanasia occurs in *Evangelium Vitae*, which will provide the focus of my discussion here. In that encyclical, John Paul notes that both abortion and euthanasia are characteristic of a "new cultural climate" which "gives crimes against life *new and - if possible - more sinister character*" (John Paul, 1995, no. 4). That sinister character derives from the justification offered for such actions as expressions of individual autonomy.[2] Thus,

> Broad sectors of public opinion justify certain crimes against life in the name of the rights of individual freedom, and on this basis they claim not only exemption from punishment but even authorization by the state, so that these things can be done with total freedom and indeed with the free assistance of health-care systems (1995, no. 4).

In the wake of this profound shift in public attitudes toward the taking of human life – ostensibly in the name of individual freedom – the integrity of the medical profession itself is under siege. Hence,

> Choices once unanimously considered criminal and rejected by the common moral sense are gradually becoming socially acceptable. Even certain sectors of the medical profession, which by its calling is directed to the defense and care of human life, are increasingly willing to carry out these acts against the human person. In this way, the very nature of the medical profession is distorted and contradicted, and the dignity of those who practice it is degraded (1995, no. 4).

This shift in cultural and political attitudes involves a

> *war of the powerful against the weak* [emphasis in original]: i.e., a life which would require greater acceptance, love, and care is considered useless, or held to be an intolerable burden, and is therefore rejected in one way or another (1995, no. 12).

Specifically in reference to assisted suicide and euthanasia, then,

> In a social and cultural context which makes it more difficult to face and accept suffering, the *temptation* becomes all the greater *to resolve the problem of suffering by eliminating it at the root*, by hastening death so that it occurs at the moment considered most suitable (1995, no. 15).

More generally stated, the above noted temptation is often linked to, or expressive of, a "certain Promethean attitude which leads people to think that they can control life and death by taking the decisions about them into their own hands" (1995, no. 15). That attitude, in turn, is linked by John Paul to a cultural and moral emphasis on freedom that is extreme in its subjectivism. Thus,

> the roots of the contradiction between the solemn affirmation of human rights and their tragic denial in practice lies in a *notion of freedom* which exalts the isolated individual in an absolute way...While it is true that the taking of life not yet born or in its final stages is sometimes marked by a mistaken sense of altruism and human compassion, it cannot be denied that such a culture of death, taken as a whole, betrays a completely individualistic concept of freedom, which ends up by becoming the freedom of "the strong" against the weak who have no choice but to submit (1995, no. 19).

This extreme version of subjectivity reflects the loss of essential moorings that characterizes the post-modern context. Within this atmosphere of unprincipled license, according to John Paul,

> freedom negates and destroys itself, and becomes a factor leading to the destruction of others, when it no longer recognizes and respects its *essential link with the truth*. When freedom, out of a desire to emancipate itself from all forms of tradition and authority, shuts out even the most obvious evidence of an objective and universal truth, which is the foundation of personal and social life, then the person ends up by no longer taking as the sole and indisputable point of reference for his own choices the truth about good and evil, but only his subjective and changeable opinion or, indeed, his selfish interest and whim (1995, no. 19).

At the "deepest roots of the struggle between the culture of life and the culture of death," one

> finds the heart of the tragedy being experienced by modern man: *the eclipse of the sense of God and of man*, typical of a social and cultural climate dominated by secularism ...Those who allow themselves to be influenced by this climate easily fall into a sad vicious circle: *when the sense of God is lost, there is also a tendency to lose the sense of man*, of his dignity and his life (1995, no. 21).

In turn, that eclipse of the sense of God "inevitably leads to a *practical materialism*, which breeds individualism, utilitarianism, and hedonism" (1995, no. 23). In such a context,

> suffering, an inescapable burden of human existence but also a factor of possible personal growth, is censored, rejected as useless, indeed opposed as an evil, always and in every way to be avoided. When it cannot be avoided and the prospect of even some future well-being vanishes, then life appears to have lost all meaning and the temptation grows in man to claim the right to suppress it (1995, no. 23).

Finally, and most importantly for the concerns of this essay, euthanasia as a response to another's suffering represents "an injustice which can never be excused":

> Even when not motivated by a selfish refusal to be burdened with the life of someone who is suffering, euthanasia must be called a *false mercy*, and indeed a disturbing perversion of mercy. True compassion leads to sharing another's pain; it does not kill the person whose suffering we cannot bear (1995, no. 66).

In the moral logic of the above arguments, John Paul appeals to a range of values: the value of a freedom linked to objective standards of right and wrong; the value of justice, in light of which euthanasia is seen as a violation of the rights of the most vulnerable in our midst under the guise of a misplaced compassion; and the value of the integrity of medicine as an oathed profession. However, the fundamental perspective at work in his discussion remains explicitly theological. The eclipse of God's purposes results in the devaluing of life, a thoroughly subjectivist account of human freedom and agency, an attitude of unconstrained technical control of the circumstances and timing of death, an impoverished understanding of the meaning of suffering, and a social climate of materialism and hedonism.

Note the stark contrast between the assumptions at work in John Paul's discussion and arguments by proponents of assisted suicide and euthanasia. In recent years, a number of secular thinkers (as well as some religious commentators), have argued that assisted suicide and voluntary euthanasia as clinical alternatives for the terminally ill are motivated by different values than those reflected in suicide as a non-clinical choice (e.g., Humphry, 1991; Maguire, 1994). At the heart of the secular debate about assisted suicide and euthanasia are two values – that of autonomy or self-determination, and a desire to avoid or end intractable pain and suffering. By contrast, John Paul's vision is informed by fundamentally different assumptions about the basic values that frame our understandings of life, death, and the facts of human pain and suffering. The notion of individual autonomy developed in secular bioethics, either as a positive value (Veatch, 1981) or as a side-constraint (Nozick, 1974), is largely at variance with John Paul's richer vision. In answer to the question in the title of the film "Whose Life is it, Anyway?", John Paul would surely answer, "God's, of course." That admission, so obvious through the eyes of a personalist faith that calls to God as Creator, Life-Giver, Redeemer, Sanctifier, Lord and Ruler, places the individual, and the potentialities of his or her freedom, within a clear horizon constituted by God's calling and purposes.

The basic context of John Paul's rejection of assisted suicide and euthanasia, then, is, in the literal sense, *radically* different. From the vantage of personalism, the individual is never one apart from community but a creature already in primary community with God and therefore with others. For John Paul, autonomy, if reduced to its secular skeleton of the "unencumbered self," emerges as a grave theological mistake, a fundamentally misguided orientation that elevates the act of choice over its content, that mistakenly views self-determination as meaningful apart from one's relation to God and His good purposes as revealed in Jesus Christ.

The second primary value of the secular debate – that intractable pain and suffering are not only to be alleviated, but if necessary eradicated by eliminating the sufferer – is equally at odds with John Paul's fundamental orientation. Granted, Christians are obliged, when possible, to respond to and alleviate the neighbor's pain and suffering as a concrete implication of *agape*. In the context of end-of-life care, a commitment to effective palliative services is a clarion implication of that neighbor love (Bresnahan, 1995). Yet, in John Paul's perspective, the self cannot be understood apart from God's purposes, which may be revealed even, and sometimes only, amidst pain and suffering. His discussion reflects the traditional Catholic understanding of the potential value to be found in suffering. For example, the Congregation for the Doctrine of the Faith, while recommending effective analgesia for "the majority of sick people," also interprets the suffering of some individuals in positive terms:

> According to Christian teaching, ... suffering, especially suffering during the last moments of life, has a special place in God's saving plan; it is in fact a sharing in Christ's passion and a union with the redeeming sacrifice which he offered in obedience to the Father's will. Therefore, one must not be surprised if some Christians prefer to moderate their use of painkillers, in order to accept voluntarily at least a part of their sufferings and thus associate themselves in a way with the sufferings of Christ crucified (O'Rourke and Boyle, 1993, p. 317).

It has sometimes been noted that such sentiments can be misconstrued in the direction of pious platitude and glib assurance, but John Paul's own discussion exhibits sensitivity and nuance. Unlike many secular approaches to pain and suffering, which often assess the latter according to a reductionist hedonism, John Paul, affirming the tradition, points to the essential mystery at the heart of the Christian experience of suffering. Although through Christ our salvation has been accomplished, and death has been overcome from the vantage of God's ultimate purposes, the penultimate realities of suffering and death are with us still. It is the "already but not yet" character of our salvation that underlies a number of realities of Christian faith and experience. Death, while having "lost its sting," remains as a physical reality, with attendant pain and suffering. But the Church, as the community of believers in the saving mysteries of Christ, mystically participates in his sufferings. Hence, "identification" with Christ's sufferings, while not a solution to the philosophical puzzles of theodicy, affords us, through faith, a sense of confidence and trust that even in such sufferings, God's purposes will not be thwarted. As Christ says to Paul in Second Corinthians, "My grace is sufficient for you, for my power is made perfect in weakness" (2 Cor. 12:9).

The account of personal freedom offered by John Paul, then, stands in marked contrast to the notion of autonomy proposed by defenders of assisted suicide. Liberty is not license, and choices directly against the good of life undercut the deeper freedom that John Paul espouses, *viz.*, the freedom to know and to choose the good. Hence, for John Paul, there is a "right not to be killed," even at one's own request, because such a choice is a denial of a basic moral truth about persons. While such a judgment will be interpreted as a restriction of individual autonomy by those for whom consent is the lynchpin of moral choice, it emerges, in John Paul's richer discussion of persons in community, as an affirmation of responsible freedom.

3. ON THE RIGHT TO BASIC MEDICAL CARE: JUSTICE-BASED ARGUMENTS FOR A POSITIVE ENTITLEMENT

John Paul's own contribution to the discussion of human rights is best understood within the general context of Catholic social teaching since Leo XIII, although there are distinctive features to his account as well. *Centesimus Annus* and *Laborem Exercens*, in their affirmation of individual rights and social solidarity, move well beyond the classicism still evident in Leo XIII's *Rerum Novarum*. On the one hand, the change in historical context since 1891 provides a greater nuance to John Paul's appraisal of the relative strengths and weaknesses of both capitalism and collectivism. While Leo's robust defense of private property is best viewed as an affirmation of individual dignity of socialism's collectivizing tendencies, his affirmation of the state's necessary role in guaranteeing the wage rights of workers remains largely "organicist" in its thrust. Leo's vision of the rights of workers, the duties of employers, and the legitimate functions of the state in assuring those rights and enforcing those duties, still presume a largely organic model of society. If pluralism is affirmed, it is a pluralism of *forms* rather than of content. Hence the respective functions of the church in relation to the modern state remain, in many ways, almost medieval in their orientation (Curran, 2002, pp. 68-71).

For John Paul, matters are far more complex. The language of rights based on individual dignity is prominent in his analysis; indeed, in his social perspective, that language assumes the priority one finds in social teaching since the papacy of John XXIII. Nonetheless, the rights that he affirms and discusses occur in the context of modern largely secular pluralism, a pluralism that in many respects emerges the object of John Paul's cultural critique. Thus, especially in *Veritatis Splendor* and *Evangelium Vitae*, he takes direct aim at the themes reviewed above in discussing euthanasia: secular freedom is deficient, indeed desiccated, for its affirmation of choice as a value in its own right, unfettered from an account of truth, *i.e.,* an understanding of what constitutes the appropriate objects of choice. By the same token, when John Paul considers positive entitlements based on considerations of human dignity, he criticizes those accounts of *political* organization that fail to acknowledge the complementarity of individual dignity and the common good, thereby undercutting the *necessary* linkage, both conceptually and practically, between both values. That complementarity derives from the Catholic tradition's insistence, as expressed systematically from Scholasticism forward, that human nature and human dignity necessarily require an account of individuals in community. Unlike those secular understandings that would interpret duties incumbent on citizens on behalf of the common good as constraints upon otherwise unfettered individual choice, John Paul's personalism situates such duties in positive terms. If human beings, by nature, are constituted to live in community, and if communities, by nature, are necessarily concerned with protecting and enhancing the freedom and well-being of individuals, both individual rights and societal obligations flow from that complementarity. Unlike libertarianism, which largely rejects the positive duties incumbent on government in respect of individual freedom, John Paul's understanding of persons in community finds an appropriate role for governmental intervention at various levels – local, national, and international – especially in the context of a modern and increasing globalized world.

The right to health care as a positive entitlement is a useful example of recent emphases in Catholic social teaching since the time of John XXIII. The conditions required by individual dignity were enumerated in a list of specific individual rights, including the right to health care, in *Pacem in Terris,*

> … we see that every man has the right to life, to bodily integrity, and to the means which are necessary and suitable for the proper development of life. These means are primarily food, clothing, shelter, rest, medical care, and finally the necessary social services. Therefore, a human being also has the right to security in cases of sickness, inability to work, widowhood, old age, unemployment, or in any other case in which he is deprived of the means of subsistence through no fault of his own (Gremillion, 1976, p. 167).

Since the time of John XXIII, Catholic social teaching has viewed access to basic health care as a positive right, *i.e.*, a justified entitlement claimable by individuals against society. The fundamental warrants for this positive right, as with other rights affirmed in the recent tradition, are those that characterize John Paul's discussion of the nature and scope of individual freedom. First is an appeal to the dignity of the person as one made in the image of God. Second is an understanding of the common good, emphasized in John Paul's social encyclical under the rubric of solidarity, which offers an organic vision of society with duties incumbent on institutions according to the purposes of society as established by God. Third is an emphasis on social justice, developed since the time of Pius XI as an extension of the traditional emphasis on the common good. Social justice is a specific regulative notion that enjoins institutions and, increasingly, governments to guarantee the basic material concomitants of human dignity. Fourth is an appeal to the principle of subsidiarity, first enunciated by Pius XI, which speaks of the intrinsic and instrumental value of meeting the basic needs of persons at the lowest or least centralized level of association and authority possible. Fifth, in recent Catholic social teaching has been an emphasis on the so-called "preferential option for the poor." For John Paul, this option is generally considered within a broader emphasis on solidarity as a broad principle binding citizens together to meet the basic needs of all, with special priority attached to meeting the needs of the poor. Sixth, though less a specific emphasis in John Paul's encyclicals, some recent commentators have emphasized distributive justice as an appeal that functions somewhat independently of the more general focus of social justice.[3]

Throughout the modern papal discussion, including John Paul's own social encyclicals, one is struck by the general nature of papal pronouncements about what individual dignity requires. That level of generality is not surprising. The encyclicals are, first and foremost, statements of theological and moral vision rather than recommendations on specific policy alternatives. They provide more than a mere statement of ideals but less than a blueprint for specific choice and action. Nonetheless, the expressly theological grounding of human dignity in recent papal documents, including the writings of John Paul, provides a different and richer context for understanding the usual appeals to liberty and equality in secular debates about the right to medical care. Again, in Catholic social teaching, although individuals are endowed with freedom, theirs is a freedom to be exercised in community. In light of the belief in God as Trinity, and in contrast to the polarities one often finds in secular debates about positive rights, Catholic social teaching, well exemplified by John Paul's writings, elevates neither liberty nor equality to a position of unchallenged priority. Rather than viewing either value as trumping the other, Catholic discussion, and particularly John Paul's language of solidarity, presents liberty and equality as mutually accommodating principles.[4]

So, too, the theologically framed appeals to the common good and to solidarity, much like the theologically framed expositions of individual dignity, provide an alternative to secular approaches that begin, and often end, with an emphasis on unfettered individualism. The language of the common good and of solidarity remains fundamentally social and institutional in its focus; it stresses human dependence and interdependence. This is not to suggest that the common good should be seen as the Roman Catholic analogue to a utilitarian calculus. Rather, in the words of John Langan, the common good

> insists on the conditions and institutions ... necessary for human cooperation and the achievement of shared objectives as decisive normative elements in the social situation, elements which individualism is both unable to account for in theory and likely to neglect in practice (Langan, 1986, p. 101).

The notion of the common good, motivated by solidarity, should not be seen, therefore, as necessarily in tension with the rights of individuals. Rather, the common good functions in two important senses: first, it is invoked to temper and correct the inequities often associated with secular individualism; and second, it incorporates guarantees for personal rights and duties. Still, in contrast to individualistic theories, a fundamentally social understanding characterizes Catholic thought. This social perspective is especially evident in recent Catholic teaching on the nature and scope of property, with implications for a number of positive rights, including the right to health care. Theologically, persons are imagers of a Trinitarian God. Practically, this suggests that the claims of individuals to resources are limited by the claims of others for the satisfaction of basic needs. In hard cases, where choice is inevitable, John Paul summarizes the conclusions of the traditions in this way:

> Christian tradition has never upheld [private property] as absolute or untouchable. On the contrary, it has always understood this right within the broader context of the right common to all to use the goods of the whole of creation: the right to private property is subordinated to the right to common use, to the fact that goods are meant for everyone (John Paul II, 1981, no. 43).

In this emphasis on property in common as a regulative notion, the common good emerges as fundamental. Individual rights and duties are seen as constitutive of the common good, but there are no absolute or unmediated claims to private ownership of property. Unlike perspectives that begin with the distinction between private and public resources as the unassailable datum from which moral analysis proceeds, the Catholic tradition does not deem individual ownership to be sacrosanct. Rather, common access according to use remains the relevant criterion according to which social arrangements and practices must be amassed, especially in the circumstances of a developed economy.

To be sure, the principle of subsidiarity functions, in the words of one commentator, as "another dimension" of the common good, because that principle recommends that responsibility be exercised "at the smallest appropriate level" (Christiansen, 1991, pp. 46-47). Subsidiarity implies that "the first responsibility in meeting human needs rests with the free and competent individual, then with the local group" (Ashley and O'Rourke, 1978, p. 132). And there is a substantive, not merely instrumental, value in the moral involvement of individuals in the interpersonal forms of association that subsidiarity commends as a functional aspect of the common good. The common good involves the basic goods of persons. Subsidiarity, as a dimension of the common good, underscores the intrinsic value of direct and immediate forms of the individual's responsibility to others.

However, there has been an increasing emphasis in recent social teaching on the necessity of governmental involvement in meeting the basic needs of persons under the rubric of social justice. As societies develop, as medicine progresses, institutions, especially at the governmental level, are morally called to mediate the claims of human dignity and to shape the content of human rights, including the right to medical care. Moreover, the appeal by John Paul and others to a "preferential option for the poor," while it may be seen as a separate appeal, can also be viewed as a practical requirement of more fundamental themes: the positive entitlements of individuals, the common good, and social justice. In service to those values, institutions are called to respond to those inequities between and among individuals that especially threaten the dignity of the most disadvantaged in society. (Whether one views the preferential option for the poor as an implication of more fundamental themes, or as having independent standing, the practical results at the level of social policy about health care and other basic goods are likely to be the same.)

As I noted above, much of the modern encyclical discussion, including John Paul's, is couched at a quite general level. Consequently, the linkage between general theological and moral claims and their concrete implications for particular issues often remains unclear. Nowhere is that imprecision more pronounced than in papal and episcopal discussions of individual rights as positive entitlements. As a case in point, I will review a recent document by the U.S. Catholic Bishops, their "Resolution on Health Care Reform," which can be seen as a specific example of the way that basic principles of Catholic social teaching have been brought to bear on health care reform in the United States. I will suggest that the conceptual unclarities and tensions found in the Resolution result from two things: first, the lack of specificity in the encyclical discussion (including John Paul's) linking the language of positive rights to claims about individual dignity; and second, inadequacies in the bishops' own efforts to bridge the gap between general norms and their implications for health care reform.

In its introduction, the resolution identifies the fundamental problems that beset present healthcare delivery: excessive costs, lack of access for many Americans, and questions about the quality of the care provided. The bishops address the document to the "Catholic community" *and* to "the leaders of our nation" (U.S. Bishops, 1993, p. 98). The document, although at times expressly theological, also attempts to speak to the broader public, primarily through its appeals to human dignity, which can be justified on both theological and non-theological grounds and by its expressions of concern about the plight of those presently underserved by the U.S. healthcare system.

The resolution appeals to the recent tradition by rooting its approach to healthcare in the three fundamental themes I discussed above. First, "[e]very person has a right to adequate healthcare. This right flows from the sanctity of human life and the dignity that belongs to all human persons, who are made in the image of God." Health care is a "basic human right, an essential safeguard of human life and dignity." Moreover, the bishops' call for reform is rooted in "the priorities of social justice and the principle of the common good." In light of these fundamental values, the bishops judge that "existing patterns of health care in the United States do not meet the minimal standard of social justice and the common good." Indeed, "the current healthcare system is so inequitable, any disparities between rich and poor and those with access and those without are so great, that it is clearly unjust" (U.S. Bishops, 1993, p. 99).

The bishops also appeal to a preferential option for the poor as a particular implication of the common good. Pointing out that the "burdens of the system are not shared equally," the bishops conclude that we must "measure our health system in terms of how it affects the weak and disadvantaged." Fundamental reform must be especially concerned with "the impact of national health policies on the poor and the vulnerable." In this context, the bishops quote with approval a recent ecumenical statement on the common good; to wit:

> More than anything else, the call to the common good is a reminder that we are one human family, whatever our differences of race, gender, ethnicity, or economic status. In our vision of the common good, a crucial moral test is how the weakest are faring. We give

> special priority to the poor and vulnerable since those with the greatest needs and burdens have first claim on our common efforts. In protecting the lives and promoting the dignity of the poor and vulnerable, we strengthen all of society (U.S. Bishops, 1993, p. 98).

As a final appeal in Section One of the Resolution, the bishops invoke the prudential notion of stewardship. The cost of present health care in the United States "strains the private economy and leaves too few resources for housing, education, and other economic and social needs." In response, "[s]tewardship demands that we address the duplication, waste and other factors that make our system so expensive" (U.S. Bishops, 1993, p. 99).

The practical focus of Section Two of the Resolution raises the same cluster of thematic concerns in more concrete fashion. In this section, the bishops set out eight practical criteria by which to judge the moral adequacy of reform proposals. The criteria are as follows: (1) respect for life, (2) priority concern for the poor, (3) universal access, (4) comprehensive benefits, (5) pluralism, (6) quality, (7) cost containment and controls, and (8) equitable financing. Given the brevity of the resolution, these criteria are not adequately developed, but as a whole, they emerge as practical expressions of the fundamental theological values that inform the Catholic discussion of health care. At the same time, because the bishops clearly intend these criteria as useful guides for assessing public policy, is important to note certain conceptual and practical tensions between and among the various criteria. For purposes of the present discussion, I will focus on the first four.

The first principle, respect for life, speaks to the need for any reform proposal to "preserve and enhance the sanctity and dignity of human life from conception to natural death." Nonetheless, sanctity of life does not, of itself, shed light upon how to proceed in cases when hard choices involving allocation of resources must be made between and among individuals, all of whom might at least marginally benefit from continued provision of care. Moreover, the relevance of this first principle to certain hard cases (including abortion and the status persistently vegetative patients) is unclear, since the grounding of "personhood" claims in these instances may depend on theological understandings of "ensoulment" or capacities to "image" God that are not available as warrants for secular policy.

The second principle – priority concern for the poor – implies, according to the resolution, that any reform proposal should be judged as to "whether it gives special priority to meeting the most pressing the healthcare needs of the poor and underserved, ensuring that they receive quality health services." Here the difference between according the preferential option for the poor independent weight or interpreting it as a particular implication of the common good may significantly affect how one assesses a given reform proposal. While it is doubtless true that working correlations between overall indices of poverty and poorer health can be drawn, it is not the case that for any particular indigent individual, health outcomes will necessarily be correlated with access to healthcare. Indeed, there are strong arguments, doubtless of a more prophetic sort, that reform form of health care as a discrete sector may be one of the least effective ways of improving the general health outcomes of the poor as a class. Overall poverty, not simply limited access to medical care, may be a far more relevant determinant of health than the bishops' second criterion suggests.

With regard to medical care, then, the preferential option for the poor will require greater attention to a definition of the "poor" for whom preferences are to be shown – those who are "generally poor," according to most indices, or those who are "medically indigent." To the extent that the unclarity persists, the relevance of the criterion for any particular patient, as compared with its relevance as a working generalization about classes of persons, will not be obvious. Indeed, for any particular patient, the first criterion, "respect for life" might suggest that general indices of poverty, in contrast to criteria relating specifically to medical need or medical indigence, may discriminate unfairly against a needy patient who would not (at least initially) qualify according to the former indices.

By contrast, if the preferential option for the poor does not have independent weight but instead is seen as a particular application or implication of the common good, then determinations of medical indigence or medical need might proceed apace with difficult judgments about basic social goods other than medical care. On this account of the preferential option, the common good might be invoked as a systematic consideration to limit the availability of resources for particular individuals as the result of prior social choices, independent of individual circumstances of need. In this scenario, one might then be able to distinguish circumstances that are admittedly unfortunate from those that are unfair.

According to the bishops' third criterion, any morally acceptable proposal must provide "universal access to comprehensive health care for every person living in the United States." A number of practical questions arise here. How shall the system be reformed in order to provide genuinely universal access to comprehensive benefits? If one relies on market mechanisms and incentives to expand access and availability, how will such competitive economics work to ensure entitlement on a universal basis? In addition, should citizenship be morally decisive in determining access to available services? The wording of this criterion would suggest that healthcare coverage be limited to those within the United States but also that all persons within U.S. borders (not only U.S. citizens) be provided care. Practically, this may pose significant difficulties, since the costs of such care, indiscriminately provided, may undercut the willingness and/or ability of taxpayers to fund a basic level of health care for all.

The fourth criterion offers a benchmark of "comprehensive benefits" for any morally acceptable reform proposal. Comprehensive benefits include those "sufficient to maintain and promote good health, to provide preventive care, to treat disease, injury, and disability appropriately and to care for persons who are chronically ill or dying." Again, the choices and tradeoffs required among various goods might well conflict with the focus on the particular individuals seemingly implied in such principles as "respect for life" or "universal access." Health care "sufficient to maintain and promote good health" for a given individual may be, in effect, a black hole, since that person's needs, according to the fourth criterion, might swallow an inordinate amount of resources to which others with lesser needs might otherwise have access. Moreover, as noted above, medical services may be fairly low indices for health.

Thus, any criteria for "healthcare" reform invoked with insufficient attention to the multifactorial nature of good health as an outcome, may be co-opted by tendencies already present in technology-driven medicine to misallocate funds that could be better spent on primary or preventive care. Finally, only "caring for" those who are chronically ill or dying might involve a number of rationing choices that, while supported on grounds of the common good or social justice, failed to comport with the individual focus of other criteria.

As I noted above, Section Two of the resolution sets forth a number of other practical criteria for assessing reform proposals, which, while interesting, emerge as fairly commonsensical in tone. By contrast, Section Three lists four "essential priorities" that the bishops urge upon readers in applying the eight assessment criteria: (1) priority concern for the poor and universal access; (2) respect for human life and human dignity; (3) pursuing the common good and preserving pluralism; and (4) restraining costs. Although each of these priorities merits scrutiny in its own right, I will focus on key tensions generated by the bishops' discussion of the first and third priorities.

In their discussion of the first priority – concern for the poor and universal access – the bishops voice strong support for "measures to ensure truly universal access and rapid steps to improve the care of the poor and underserved." In light of that commitment, they "do not support a two-tiered health system since separate healthcare coverage for the poor usually results in poor health care. Linking the health care poor and working class families to the health care of those with greater resources is probably the best assurance of comprehensive benefits and quality care." Nonetheless, in discussing their third priority – preserving pluralism – the bishops emphasize the following:

> We believe the debate can be advanced by continuing focus on the common good and a healthy respect for genuine pluralism. A reformed system must encourage the creative and renewed involvement of both the public and private sectors ... It must also respect the religious and ethical values of both individuals and institutions involved in the health care system (U.S. Bishops, 1993, p. 101).

While it is true that both priorities, as normative generalizations, may help frame policy discussions on healthcare reform, the potential for tensions between a commitment to a single-tiered system and a respect for pluralism of individuals and institutions is, as a matter of practical policy choice, enormous. Moreover, in light of the theological convictions basic to the broader Catholic discussion of health care, considerations of individual liberty and dignity, as well as of the common good, might reasonably lead to a different practical conclusion; viz., that two tiers of health care delivery are both morally appropriate and practically preferable, so long as universal access to comprehensive basic care is assured. Consider, for example, basic education as a useful analogy to health care: a tax-based commitment to education for all citizens does not prevent individual parents from paying for alternative basic schooling for their children. It is not obvious, at least without a great deal more argument then the bishops provide, that health care should be viewed differently. Nor is it obvious, in light of the general theological warrants I analyzed earlier, that the common good would necessarily dictate the priority of equality over liberty in the delivery of health care.

The bishops' tendency to ignore or under emphasize conceptual and practical tensions among the normative criteria and policy priorities they develop exemplifies tensions to be found in Catholic social teaching more generally, including the social encyclicals of John Paul. First, the bishops (and pontiffs) often conflate hortatory and pragmatic concerns in the discussion. Second, in light of that conflation of different sorts of moral discourse, a great deal more by way of careful, practically oriented argument will be necessary before such assessment criteria and normative priorities can be considered significant contributions to policy debates on specific topics. Third, the conceptual and practical tensions in the papal and episcopal discussions will require their readers to identify the various modes of moral discourse at work in their recommendations in order to appreciate the different ways that Catholic themes and principles may be relevant to the secular discussion. In light of such uncertainties, I now turn, in closing, to a summary overview of several conceptual and practical tensions posed by Catholic social teaching (including the writings of John Paul) that merit further scholarly scrutiny.

4. CONCLUDING REFLECTIONS

The general themes of Catholic social teaching, as they relate to the structures and governance of pluralistic societies, exhibit a number of tensions. As I noted in analyzing the Bishops' Resolution, those difficulties emerge at the level of policy choice in attempting to translate general directives into concrete options. At one level, perhaps, that lack of specificity is unsurprising, given the broad character of encyclical documents which provide the context for the bishops' more focused deliberations. As Charles Curran observes,

> The absence of controversy about the response to authoritative papal social teaching seems to derive from four different sources. First, the social teachings themselves are quite general; disputes tend to arise on more specific issues. Second, the hierarchical magisterium itself has not made an issue of dissent in this area as it has in the area of sexual morality. Third, the hierarchical magisterium in social matters has been more open to change and development on many issues, such as the role of government, the forms of government, religious liberty, and human rights. Fourth, the greater fascination with sexual issues on the part of the church and society has brought more attention to these issues than to the issues involved in social teaching (Curran, 2002, p. 113).

Such practical reservations aside, there remain basic unclarities that that make it difficult to assess fully the cogency of Catholic social teaching as a useful general framework for reflection. Three such issues seem especially deserving of sustained scholarly scrutiny. First is the question of how various moral appeals are to be theoretically integrated. There are a mixture of appeals in Catholic social teaching generally, and in John Paul's corpus as well – to natural law, to social duties, to solidarity, to individual rights. But there remains a notable lack of attention to the tensions between and among such appeals. Natural law is subject to criticism, both within and without Catholicism, for its failure to provide an unambiguous basis for insight or action of particular policies and practices, beyond the most general level of moral agreement. Even on matters of life and death – for example, abortion, assisted suicide, and euthanasia – the putative obviousness of such natural law appeals is not persuasive to many fellow Christians nor to those of self-consciously secular allegiances. Unless one is prepared to impute ignorance or bad faith to those who fail to draw "obvious" conclusions in such instances, or to admit that natural law reasoning, even about supposedly obvious matters, must be illuminated by theological sources of insight, natural law will remain a questionable basis for general appeals to either broad-based social duties or to individual rights. (Of course, to adopt the second strategy, i.e., to admit the need for implicit or explicit theological commitments, would seem a Pyrrhic victory, since the putative obviousness of natural law, given that interpretation, is again undercut.)

Second, there are questions about how particular warrants are meant to function in different contexts of moral appeal and discourse. James Gustafson has recently analyzed four such modes – what he calls ethical discourse, prophetic discourse, narrative discourse, and policy discourse. Each of these modes, Gustafson suggests, is at some level necessary to the moral deliberations of particular communities and society at large, but none is, of itself, sufficient (Gustafson, 1990). In assessing various arguments, one must attend to the warrants at work in each mode of discourse and determine the different degree of authoritativeness associate with each. It is fair to conclude that Catholic social teaching, in both its papal and episcopal expressions, has been characteristically imprecise in its various moral appeals. Narrative ethics, aimed at the Christian community often functions side by side with recommendations for general social policy, with little acknowledgement of the theoretical (and practical) tensions between those different modes of discourse.

Finally, for David Denz and other critics, the central task of Catholic social teaching is to be more specific about the warrants and underpinnings of its own positions. Since the time of Leo XIII, the encyclical literature, as well as pastoral and policy statements on particular issues by national bodies of bishops, have dramatically moved the Church toward sustained engagement with issues of economics, politics, peace, and world order. At its best, social teaching has affirmed the "primacy of human dignity and human solidarity as correctives to mere technocratic understandings of the economy and politics" (Coleman, 1991, p. 3). Nevertheless, Denz is especially critical of the social teaching tradition on a key point, what he calls the "derivation problem." Since the papacy of John XXIII, social teaching has emphasized individual rights based on considerations of dignity. John Paul's writings have both echoed and extended that general emphasis. To be sure, rights may, indeed, be appropriately viewed as correlative claims to earlier formulations of natural law-based duties. Yet there remains a troubling imprecision in the papal and Episcopal discussion of such rights, especially those involving positive entitlements. Consequently, both the nature and scope of such rights therefore remain at issue. As Denz trenchantly observes, "we need to know how dignity generates rights, both for the sake of determining the extent of those rights and for the light this may shed on how those rights constrain or balance one another." Unless greater attention is paid to clarifying the linkage between dignity and specific rights, including the right to health care, the central assertions about rights in the tradition of social teaching may emerge as a Catholic version of "manifesto rights," prescinding "so far from actual institutional life that [they cease] to suggest a plausible institutional embodiment" (Denz, 2000, p. 255).

Rice University
Houston, Texas

NOTES

1 According to John Conley, when discussing the plight of Cain in *Evangelium Vitae*, John Paul's exegesis reveals the influence of phenomenology: "In discerning the universal values and practical consequences of Cain's resort to murder, the Pope manifests the objective 'essence' of the human person faced with the drive toward violence. This unveiling of the objective essence of humanity, especially the objectivity of its key moral values and attitudes, is central to the phenomenological project of Max Scheler, the phenomenologist studied by the Pope in his doctoral dissertation, and other Catholic participants in 'Munich phenomenology,' such as Dietrich von Hildenbrand. Further, the Pope's movement of argument closely follows the method of phenomenological analysis, which moves from the more superficial to the more profound levels of the problematic subject of analysis. The essence of the subject is gradually unveiled through successive layers of meditation upon the multiple appearances of the subject" (Conley, 1997, p. 6).

2 Efforts to legalize physician-assisted suicide or euthanasia may be driven by various motives. Thus *Evangelium Vitae* speaks about the personal and cultural dimensions of the crisis it portrays: "Decisions that go against life sometimes arise from difficult or even tragic situations of profound suffering, loneliness, a total lack of economic prospects, depression, and anxiety about the future. Such circumstances can mitigate even to a notable degree subjective responsibility and the consequent culpability of those who makes these choices which themselves are evil. But today the problem goes far beyond the necessary recognition of these personal situations. It is a problem which exists at the cultural, social, and political level, where it reveals its more sinister and disturbing aspect in the tendency, ever more widely shared, to interpret the above crimes against life as *legitimate expressions of individual freedom, to be acknowledged and protected as actual rights"* [emphasis in the original] (1995, no. 34).

3 As Philip Keane observes, "... reform of the structures of society so that society can more effectively deliver health care still depends on society having a clearer focus on just what health care goods it ought to be delivering to people. Thus, while the social structures question, like the equality question, is a pivotal aspect of all justice including health care justice, my judgment is that the distribution question (i.e., what health care benefits must we provide?) is the most central of all the justice questions which relate to health care ... If we focus too much on social justice, the risk is that we will emphasize the structures necessary to furnish health care instead of first focusing on the human need of real persons to have real health care crises adequately addressed. Such emphasis on structures without a prior commitment to genuine human needs can raise all the fears of complex bureaucracies without really substantial goals" (Keane, pp. 138-139).

4 "... Our Trinitarian theology becomes a radical challenge to community. How can Christians say they believe in God if they are unwilling to put together structures that build human community and meet fundamental human needs? How can someone *really* claim to believe in the triune God and not feel a sense of outrage about the quarter of the U.S. population which lacks or is inadequately supplied with such a basic good as health care coverage? If we believe in the triune God as the very ground of community, the problem of our health care system is not just an ethical or an economic or political problem. The problem is ultimately a religious or theological problem" (Keane, p. 105).

REFERENCES

Ashley, B. and K. O'Rourke (1978). *Health Care Ethics*. St. Louis, Missouri: Catholic Hospital Association.

Bresnahan, J. (1995). 'Observations on the Rejection of Physician-Assisted Suicide: A Roman Catholic Perspective.' *Christian Bioethics* 1(3) 256-284.

Christiansen, D. (1991). 'The Great Divide.' *Linacre Quarterly* 58 (May) 40-50.

Coleman, J. (Ed.) (1999). *One Hundred Years of Catholic Social Thought: Celebration and Challenge*. New York: Orbis Books.

Curran, C. (2002). *Catholic Social Teaching 1891-Present*. Washington, D.C.: Georgetown University Press.

Denz, D. (2000). 'Catholic Social Teaching and Health Care: Some Reservations.' *Christian Bioethics*. 6(3) 251-266.

Donders, J. (Ed.) (1996). *John Paul II: The Encyclicals in Everyday Language*. Maryknoll, New York: Orbis

Books.
Gremillion, J. (Ed.) (1976). *The Gospel of Peace and Justice: Catholic Social Teaching Since Pope John*. Maryknoll, New York: Orbis Books.
Gustafson, J. (1990). 'Moral Discourse about Medicine: A Variety of Forms.' *The Journal of Medicine and Philosophy* 15(2) 125-142.
Humphry, D. (1991). *Final Exit: The Practicalities of Self-Deliverance and Assisted Suicide for The Dying*. Eugene, Oregon: The Hemlock Society.
John Paul II (1981). *Laborem Excercens*. Washington, D.C.: United States Catholic Conference Office of Publishing
Services.
John Paul II (1993). *Veritatis Splendor*. Boston: Daughters of St. Paul.
John Paul II (1995). *Evangelium Vitae*. Boston: Daughters of St. Paul.
Langan, J. (1986). 'Common Good.' in J. Childress and J. Macquarrie (Eds.), *The Westminster Dictionary of Christian Ethics*. Philadelphia: Westminster Press.
Maguire, D. (1994). 'Moral Dominion Over Dying: The Case for Mercy Death.' *Bioethics Forum* 10(2) 20-23.
Nozick, R. (1974). *Anarchy, State, and Utopia*. New York: Basic Books.
O'Brien, D. and T. Shannon (Eds.) (1992). *Catholic Social Thought: The Documentary Heritage*. Maryknoll, New York: Orbis Books.
U.S. Bishops (1993). 'Resolution on Health Care Reform.' *Origins*. 23, 98-102.
Veatch, R. (1981). *A Theory of Medical Ethics*. New York: Basic Books.

CHAPTER TEN

JOHN F. CROSBY

KAROL WOJTYLA ON TREATING PATIENTS AS PERSONS

1. INTRODUCTION

As John Paul II was recovering in the hospital in Rome following the attempt that was made on his life in May of 1981, he spoke with his doctors about the dangers that go with being a patient. On one occasion, according to Andre Frossard, who apparently heard this directly from John Paul himself, he

> explained to them how the patient, in danger of losing his subjectivity, had to fight constantly to regain it and once more become "the subject of his illness" instead of simply remaining "the object of treatment." He pointed out that the doctors are certainly not responsible for this state of affairs...but that they ought to be aware of the danger and of the efforts which the patient is obliged to make to regain control of himself. This problem of the transformation of the individual into a thing occurs everywhere in the realm of social relations. According to John Paul II it is one of the biggest problems of philosophy – and one of the most serious problems in the modern world (Frossard, 1989, p. 249).

Those who are familiar with the mind of John Paul are not surprised to read this report. We immediately recognize as vintage Wojtyla these personalist reflections on being a patient. We recognize some constantly recurring themes in John Paul's teaching, and we find it entirely natural for them to appear in conversation with his physicians. If we are acquainted with his early philosophical writings we readily recognize Wojtyla's "acting person," endangered in his "integration," struggling to preserve his "self-possession."

Now I propose to examine Wojtyla's contrast between being "the subject of illness" and being "an object of treatment," and to examine it in the light of his personalism, especially as this is expressed in his pre-papal philosophical papers and books. Thus it will usually be more correct for me to speak of Karol Wojtyla than to speak of John Paul, even if the personalism that I will be studying has entered intimately into all of his papal teachings. Though Wojtyla has never elaborated on the ethics of the physician-patient relation, as far as I know, we can gather from his personalism how he might have elaborated on it. But the point is not just to guess at what he would have said on this subject but rather to be empowered by his personalism to think for ourselves about what it means for physicians to take their patients seriously as persons.

C. Tollefsen (Ed.), *John Paul II's Contribution to Catholic Bioethics*, pp. 151–168.

2. SUBJECTIVITY

We notice the talk of "subjectivity" in the report that Frossard gives of Wojtyla's conversation with his doctors. This is in fact our natural point of departure: Wojtyla's understanding of personal subjectivity.

St. Augustine is well-known for describing the danger of losing ourselves in the world outside and for admonishing us to turn within, to enter into the "inner man." He explores the interiority of man like no one before him did. Now Karol Wojtyla, too, is fascinated with the interiority of persons. He announces one of the great themes of his personalist philosophy when he writes: "we can say that the person as a subject is distinguished from even the most advanced animals by a specific inner self, an inner life, characteristic only of persons. It is impossible to speak of the inner life of animals" (Wojtyla, 1981, p. 22). This interiority of which Wojtyla speaks is nothing other than personal subjectivity; the two terms are interchangeable for him, though he clearly prefers the term subjectivity.

Let us consider the way in which we know the world around us – plants, rocks, clouds, stars, houses, animals, other human beings. We know them all as *objects* of our experience, that is, as standing in front of us, as outside of us. But ourselves we know in a fundamentally different way; we do not just stand in front of ourselves, looking at ourselves from outside of ourselves, but we first experience ourselves in the more intimate way of *being present to ourselves*, that is, we first experience ourselves not from without but from within, not as object but as subject, not as something presented to us but as a subject that is present to itself.[1] Now this self-presence is the interiority, or subjectivity, of a human person.

A rock has an inner side, which is revealed when the rock is split, but this inner side has nothing to do with interiority. For the only way the inside of the rock can be related to experience is as object of someone's experience, it does not experience itself from within itself and hence has no subjectivity. The inside of the rock is as external as the outer surface of it; the rock is incapable of that dimension of being that we call interiority, or subjectivity.

This interior self-presence, in which each person dwells with himself, is easy to overlook. When we think about something, give attention to it, or talk about it, we put it in front of ourselves, and so it comes natural to think that this is the way we experience even ourselves. Of course, we can make an object of ourselves and of our inner life, as when we tell someone about our feelings, but our primary experience of ourselves is not from without as object, but from within as subject, and so this self-experience is in a way hidden from our view.

Now Wojtyla teaches that we must take account of our subjectivity if we are to do justice to ourselves as persons. He says that for too long philosophy tried to understand man without sufficiently consulting the evidence of subjectivity. Even Aristotle looked at man mainly from the outside, examining man in the same way he examined plants and animals. He used the same categories for explaining man and the other beings in nature, categories such as substance or matter/form. He was with his "cosmological" approach to man, as Wojtyla calls it, still able to see that man ranks higher than plants and animals, but he was not able to do justice to man as person. Only the exploration of subjectivity that begins with St. Augustine discloses the mystery of each human being as person.[2]

Of course, Wojtyla does not propose that the personalist perspective should replace the cosmological perspective. Man is not exclusively a being of subjectivity. We will later make mention of the rich philosophy of personal embodiment in Wojtyla. The cosmological perspective retains for him its own truth; the task is to keep it from being our only perspective and to enrich it with the experience of subjectivity.

Here are some connections between subjectivity and personhood that are discussed in the writings of Karol Wojtyla; they will enable us to understand the connection more deeply.

1. If we look at the sexual union of man and woman from the outside, that is, from a cosmological perspective, we notice primarily the procreative power of it. As a result we are struck by the similarity between the sexual union among human beings and the sexual union among certain sub-human animals, which after all also reproduce by means of the coupling of male and female. One sees why Wojtyla says that the cosmological perspective tends to "reduce man to the world," that is, to stress the continuity of human and non-human beings.

Let us now bring in the factor of subjectivity. Let us ask how man and woman experience their marital intimacy from within, as only they can experience it. We find that they experience something that has no counterpart among the animals and is entirely distinct from the procreative potential of their union; they *will to make a gift of themselves to each other* in their spousal intimacy. This self-donation throws into relief the discontinuity rather than the continuity between human and sub-human animal sexuality. But self-donation is not apparent to one looking in on man and woman from the outside; only one who knows something of the interiority of man and woman can find this entirely new dimension of sexual union. We could say that spousal self-donation is cosmologically invisible; only those who dwell in the world of subjectivity can find it.

Here is the point that Karol Wojtyla wants to make as a personalist philosopher: we get a more personalist understanding of the sexual union of man and woman by looking into their subjectivity and bringing to light their will to self-donation. In giving myself to another I do not just participate in a cycle of nature and resemble other animals, but I also perform an eminently personal act. If we remain content with an exclusively cosmological perspective, seeing nothing more in sexual union than its procreative potential, then our understanding of it is not yet a truly personalist understanding.

It goes without saying that Wojtyla does not want the personalist focus on spousal self-donation to replace the focus on reproduction. He is after all well known for affirming the inseparability of the unitive and the procreative meanings of the marital act, and in fact he has made a great effort throughout his life to give a convincing account of this inseparability. What Wojtyla aims at is a personalist enrichment of the traditional understanding of the marital act, an enrichment that is gained by attending to spousal subjectivity.

By the way, we have here the reason for the principle of selection that Wojtyla uses in his profound commentary on the passage in Genesis 1 dealing with the creation of man and woman. He remarks that there are in fact two different accounts of the creation of man and woman in Genesis 1 and that one of them is more "subjective" and the other more "cosmological." He means that one of them stresses the self-experience of Adam and Eve – the *solitude* of Adam before the creation of Eve, the *shame* of Adam and Eve after their fall – more than the other. In other words, one takes more account of the subjectivity of Adam and Eve than the other does. He focuses most of his commentary on the more subjective or interior passage because it lends more support to his personalist interpretation of man and woman (John Paul II, 1997, pp. 29-32).

2. In John Paul's encyclical, "On Human Work" (*Laborem Exercens*, 1981), we have still further evidence of his concern with subjectivity. He makes a great point of distinguishing the objective from the subjective aspects of human work (nos. 5 and 6). It is not enough to think of work in terms of its productivity; we have also and above all to think of it in terms of the human flourishing of those who perform the work. Man is not made for work, but work exists for man: this is Wojtyla's way of calling attention back to the subjectivity of work. He insists that the subjective impact on the humanity of the workers is a far more important aspect of work than the objective productivity and efficiency of work. He draws the conclusion that certain rankings that were traditionally made with regard to kinds of work are in a way relativized when one attends sufficiently to the subjective side of work. Though the product of one kind of work may be of greater value than that of another kind of work, the fact that in each case it is a person who performs the work imparts a fundamental dignity to the work. The real differences in value between this and that kind of work are not effaced, but the fact that it is a person working in each kind has a sort of equalizing effect on one's final assessment of rank.

To stress the objective aspect of work at the expense of the subjective would be for Wojtyla analogous to stressing the procreative aspect of the marital act at the expense of the love aspect.

3. We can find the concern with subjectivity at an even more fundamental level of Wojtyla's thought. In his philosophy of freedom he says that philosophers have often given an account of freedom that fails to get at the real soul of a free act. They have limited themselves to goods that we act to realize or to protect, and to the evils we act to avert or to destroy, and have overlooked the self-determination at the heart of freedom. Suppose I generously help a person in need. The freedom that I exercise is not accounted for adequately by referring to the good that I act to promote, namely the good of that person's life and health; it is not accounted for adequately by referring to this outward thrust, or "horizontal transcendence" of my act, as Wojtyla calls it. One takes the full measure of my freedom only by also taking account of the way in which I determine myself in acting to help the needy person. In generously willing the good of the other, I also have to do with myself, I will to be a certain kind of self, I determine myself to be such a self. It is not that I make myself an object of my freedom, putting myself in front of myself in the manner described above; no, I determine myself from within myself, as only I (and no other) can determine myself. It is just because my self-determination is embedded within my subjectivity, that it easily escapes the analytic attention of philosophers. Wojtyla wants to retrieve it from the neglect it often suffers and to integrate it into the "transcendence towards good" which he thinks has predominated in traditional discussions of freedom. He even wants to say that in a certain sense freedom is above all self-determination. Once again we find him turning away from what it outside of us and is given over against us (or at least

turning away from a onesided preoccupation with it), and turning towards what is experienced first of all in the self-presence of the person. And once again he does this for the sake of vindicating the personal, for he thinks that man is understood as person only when he is understood as one who determines himself.

4. His turn to subjectivity does not stop here, calling particular attention to the moment of self-determination in freedom. He is led deeper into subjectivity when he asks how the person must be constituted so as to be able to determine himself. He answers that persons first of all possess themselves, or belong to themselves, and only for this reason are in a position to determine themselves. Persons are not just "there," as if they were things, or even spiritual things; they exist rather in the reflexivity of possessing themselves, or belonging to themselves. Indeed, this reflexivity is constitutive of persons, it is the very heart of personal being. This is why the approach to human beings in terms of interiority or subjectivity is so revealing of them as persons. We can readily discern this reflexivity in the consciousness of persons, who experience themselves as *present to themselves*, as we saw; self-presence is simply the first way in which persons live their reflexivity; it is the beginning of personal subjectivity. We understand, then, why Wojtyla finds it too cosmological to think of persons merely as substances, even as composite body-soul substances; he wants to gain the more personalist perspective that comes from turning within and discovering the self-possession which gives the body-soul substance of man its distinctly personal form.[3]

One can see how self-possession underlies self-determination by returning to the case of spousal self-donation. Persons can give themselves away only because they first belong to themselves; they can dispose over themselves in the way of spousal self-donation only because they first possess themselves in the way of persons. If we do not think that the conscious sub-personal animals do anything like give themselves away (or throw themselves away self-destructively, for that matter), it is because we do not think that they belong to themselves and are handed over to themselves.

We are now in a position to understand one of the most fundamental distinctions in Wojtyla's personalism, the distinction between person and nature.[4] By person he means the active principle of acting through oneself, and by nature he means the passive principle of being acted upon or of undergoing. He wants to say that human beings live according to person in themselves only insofar as they act through themselves. Aquinas expressed the same thought when he said, with the wonderful precision of the Latin language, that persons "non aguntur sed per se agunt." What exactly does it mean to "act through oneself"? Wojtyla answers in subjective terms, saying that we act through ourselves when we live out of our self-possession, or exercise our self-possession. We begin to act through ourselves already in being present to ourselves, and we eminently act through ourselves in determining ourselves. And here is the interiority of which we have been speaking; only that which is born of our self-possession is fully interior to us; if something only befalls us, and even if it is consciously experienced, it does not belong to our interiority.

Some readers of Wojtyla are at first perplexed at him using "subjective" in so positive a sense. They are used to "subjective" being used as a term of opprobrium and "objective" being used approvingly. But among many philosophers "subjective" and "subjectivity" take on a very positive sense when they are used to express personal interiority. When Wojtyla speaks with enthusiasm about the subjectivity of persons, as he often does, he is not making any concessions to "subjectivism," which is the destructive philosophy that takes things to be nothing more than I experience them to be. Wojtyla does not mean that a person is nothing more than his self-experience, nor does he mean that reality is nothing more than a person takes it to be; he only means that self-presence reveals like nothing else the mystery of each human being as person.

3. INFORMED CONSENT

When then John Paul speaks of a patient staying intact as a "subject of illness," he means staying intact as person, he means resisting the dangers of depersonalization that are inherent in being treated for illness. Now one way in which a doctor can depersonalize a patient is by undertaking some procedure on the patient without first seeking his or her free and informed consent. I assume in this discussion a patient who is competent, one capable of giving informed consent. And I assume a doctor who intends nothing other than to restore the patient to health, that is, a doctor who is not in danger of compromising his commitment to his patient as a result of having other interests, such as research interests. Finally, I assume a procedure that involves some risk for the patient and that has some alternatives. Now if the doctor undertakes such a procedure without getting informed consent, in the conviction that he alone knows what is in the best interests of the patient, then he degrades the patient to what Wojtyla calls "an object of treatment" and ignores the patient as person. He makes the patient passively endure the treatment and deprives him of any opportunity of acting through himself in relation to it. The self-possession of the patient, which might have been engaged, is bypassed.

The point is not that the patient is more liable to be harmed when his informed consent is not sought, as if the patient, when consulted, might catch some mistake of the doctor. It is true that an alert patient can often – and more often than one might think – detect mistakes of the physician and in this way avert harm to himself. This alertness of the patient is doubly important if the physician brings some research interest to a proposed medical procedure; in this case the informed consent of the patient provides a bulwark against a natural tendency of the physician to expose the patient to too much risk. But according to Wojtyla's personalism the primary point of seeking the informed consent of the patient is neither to minimize harms to the patient nor to maximize benefits for the patient: it is to show respect for the patient as person, to help him or her to remain intact as the "subject of the illness."

Personalist authors, Wojtyla included, usually explain the violating of persons as a using of persons. Thus when Wojtyla rethinks the issues of sexual morality in his early work, *Love and Responsibility*, he almost always explains wrong sexual behavior in terms of men and women selfishly using each other for their gratification. But here in the setting of medical practice we see how contempt can be shown for persons without any using of them. If a paternalistic doctor really wants what is best for his patients and if he competently provides for them the best possible medical care, he can hardly be said to be using them; yet if he fails to seek their informed consent he shows contempt for them as persons and undermines their personal subjectivity, as Wojtyla puts it. It is different if the doctor has strong research interests; such a doctor will be tempted to use his or her patients as a means for getting answers to his questions and advancing his research. But the strictly paternalistic doctor, unencumbered by research interests, singlemindedly committed to the health of his patients, may be beyond any suspicion of using his patients; and yet in his paternalism he shows disrespect for his patients as persons.

And so Wojtyla's personalism coheres entirely with the growing commitment to seeking the informed consent of patients for their treatment. Wojtyla would welcome this recent development in health care ethics. But he avoids the excesses of certain authors. In particular, he does not think that the informed consent of a person necessarily suffices to make the acts of that person morally well ordered. An act of informed consent to euthanasia (which would make for voluntary euthanasia) remains morally disordered. And the reason is this, that the belonging of human persons to themselves is not the "last word" about their being. If it is true that as persons they belong to themselves, it is also true that as creaturely persons they belong to God, in and through whom they exist. Their belonging to God does not abolish their belonging to themselves, but it does qualify it, modifying and restricting the way they belong to themselves. An act of self-determination that runs afoul of their belonging to God does not befit who they are. It is commonly held among theists that taking one's life, or enlisting the help of others in ending one's life, runs afoul of the belonging of the creature to God. Though one exercises one's belonging to oneself, one acts at odds with the creature that one is. Important as it is to be alive as "acting person" in relation to one's medical care and not simply to endure passively this care being paternalistically dispensed by a physician, more is required for acting as a morally well-ordered person. We might say that for Wojtyla the exercise of one's self-possession is a necessary but not a sufficient condition for the flourishing of a suffering person. The exercise of one's self-possession secures one against becoming an "object of treatment," but does not secure one against other disorders of personal existence. One sees, then, how Wojtyla distances himself from certain affirmations of autonomy made by some proponents of informed consent.

And in another respect Wojtyla distinguishes himself from certain bioethicists who plead the cause of informed consent. He does not simply affirm the imperative to seek the informed consent of patients; he traces this imperative back to the personhood of the patients. As we have seen, he shows what it is about persons that requires physicians to seek their informed consent. In this way he gives an unusually thorough grounding of the imperative. One has only to look at the discussion of informed consent in the well-known textbook of bioethics by Beauchamp and Childress; these authors make no attempt to establish any connection between the imperative to seek informed consent and the personhood of patients, nor do they attempt any other comparably fundamental grounding of the imperative (Beauchamp and Childress, 2001, Ch. 3).

We can also discern a remarkable theological dimension to Wojtyla's thought on informed consent; he holds that God waits for our "informed consent" before asking something from us: "...by giving man an intelligent and free nature, he has thereby ordained that each man alone will decide for himself the ends of his activity, and not be a blind tool of someone else's ends. Therefore, if God intends to direct man towards certain goals, he allows him, to begin with, to know those goals, so that he may make them his own and strive towards them independently. ...God allows man to learn His supernatural ends, but the decision to strive towards an end, the choice of course, is left to man's free will. God does not redeem man against his will" (Wojtyla, 1981, p. 27). If asked for biblical particulars Wojtyla would surely begin by pointing to the fiat of Mary, given only after the angel had intimated to her the divine plan of salvation (Luke 1:26-38). To connect this theological consideration with our bioethical discussion, we have only to observe that if God does not redeem man against his will, a physician should not treat a patient against the patient's will.

But by noting Wojtyla's approval of the imperative to seek and to receive the informed consent of patients we have only scratched the surface of what it takes, according to his personalism, to treat patients as persons.

4. AGAINST PATERNALISM IN HEALTH CARE

Mention was made above of paternalism in health care. A certain paternalism certainly shows itself in failing to seek the informed consent of a patient for medical treatment, but it shows itself in other ways too. As I use the term in this essay it refers to something much more encompassing than failing to seek informed consent, it refers to a style of delivering health care whereby a physician approaches his patients in the manner in which a parent approaches a child. He does not need to consult the patient or listen extensively to the patient, any more than a parent consults a child about how it would like to be educated, or where it would like to live. The parent knows better how to educate the child and so the parent decides on his own about the education of the child, and in a similar way the paternalistic physician knows better what is best for the patient and so he decides largely on his own about the medical care of the patient.

In the light of his personalism Wojtyla would find this paternalism highly suspect. If a patient wants to be heard, to be taken seriously, to be co-responsible with his physician for his health care, then the physician who refuses to listen and to deliberate with the patient, degrades the patient to an "object of treatment." But even if a patient – and I continue to assume a competent patient – wants none of this and wants instead to be provided for in a paternalistic way, the paternalism remains, from the point of view of Wojtyla's personalism, entirely suspect. Such patients – and there are very many such patients – are evading the responsibility for themselves that goes with existing as self-possessing persons. They are implicated in an interpersonal disorder that is akin to what Wojtyla in one place calls "conformism."[5] They are passively enduring medical treatment when they might be acting through themselves and sharing responsibility for it.

On Wojtyla's principles a well-ordered relation between physician and patient begins with the physician not only examining the patient and reading test results about the patient, but also with listening to the patient. Perhaps the first thing to say is that the physician must not give a strictly causal interpretation to all the utterances of the patient, for there is no subjectivity in a being that is causally completely determined. Only if the physician *understands* what *motivates* the patient can the subjectivity of the patient begin to emerge for the physician.

There is a wonderful passage in Guardini in which he describes the process of extricating myself from a too objectifying relation to another person (1965, pp. 127-128). Guardini says that I must not move closer to the other, as we might at first think, but that I must rather *pull back from the other and must grant the other the space in which he can appear before me as a person living out of his own center.* This is what the physician must do – listen to the patient to the point of discerning the personal center out of which he lives, or in other words, to discern the patient as the "subject of the illness." Once this other center of personal life has emerged for the physician, it becomes natural for him to let the patient have a voice in the deliberations about his health care and to let him be co-responsible for the decisions that are made about his health care. But it can happen that the physician is so preoccupied with providing for the well-being of the patient that he hears only his own voice and is aware only of his own center of acting, and then the patient is nothing but the "object of treatment." The treatment may be beneficent and effective, given without any ulterior motives, but the patient is still only the object of it, and so is not really taken as person.

We will discuss below, toward the end of this paper, a particular kind of listening that a personalist physician must not neglect, namely listening to the patient's account of how she experiences her own body and her bodily illness. We will try to give the personalist reason why the patient's self-experience can be an important source for the physician in making his diagnosis – why he must not dismiss it as merely subjective (in the negative sense) but must take it very seriously.

After letting the patient emerge as personal subject, the personalist physician acts to empower the patient to co-responsibility for his treatment. He does this, first of all, by truthfully informing him of his condition and prospects. It would in almost all cases be paternalism if the physician were to withhold bad news so as to spare the patient stress and anxiety.[6] The patient cannot determine himself in relation to his illness if he does not know the truth about it.[7] The physician also presents different treatment strategies that are, in his medical judgment, open to the patient. Now it is all-important to understand that the co-responsibility which concerns Wojtyla has nothing to do with a strictly medical co-responsibility, as when two doctors share responsibility for the care of the same patient. The personalism of Wojtyla would not require that the patient inform himself about his medical condition so as to make himself as far as possible into a kind of junior colleague of his physician. No, the patient provides something very different from medical expertise of the kind that a second physician might provide; he gives voice to the subjectivity of the suffering person, to the world in which the suffering person lives. For example, if the patient is suffering from terminal cancer, he may face the choice of undergoing aggressive treatment that would prolong his life even into a state of extreme debility, or of undergoing merely palliative care that would let death come sooner but before the onset of the extreme debility. The choice cannot be made on strictly medical grounds, for both choices are equally good, or equally bad. The patient should in such a case be given the space in which to show by his choice what kind of person he is, what his fundamental values are. It would be the worst kind of paternalism if the physician were to preempt this choice and try to make it for the patient. The most that the physician might do is to say to the patient who is having difficulty deciding, "Being the kind of person I know you to be, I think you would probably be happier with this treatment than with that treatment." But the physician must refrain from saying in effect, "If I were in your circumstances I would choose this treatment," and to say this as if he were giving good advice to the patient.

Grisez expresses the point well when he says, "patients have important concerns other than life and health – their relationships with God and with other people, family and work responsibilities, and so on. They, not you [the physician], can and must decide how to integrate these other concerns, which you do not share with them, with their concern about survival and health, which you do share."[8] If the health of a patient were the supreme good for a human being, or if it were the only good, then the doctor would be right to be peremptory in making recommendations, since he has by far the greater competence in securing the good of health. But since a patient is capable of many other goods besides health, some of them greater goods than health, and since he knows himself in relation to these other goods in a "subjective" way in which the doctor cannot know him, and since he is the one who participates in these goods and not the doctor, it follows that his judgment about his existence as a whole will sometimes trump the medical recommendations of the doctor. Suppose that a patient with a serious heart condition is called to the bedside of a dying parent. To reach the place where the parent lies dying the patient must travel at altitudes that will cause serious stress to his heart. The doctor has no right to try to dissuade the patient from making the trip; the patient alone knows whether this is a risk that he should take. The serious medical concern of the doctor has force only *prima facie* for the patient; it can be overruled and set aside by still more serious non-medical concerns of the patient. If the patient sets out on the risky trip the physician's task is to find ways to minimize the risk to the patient.[9]

One might think that the anti-paternalism that I claim to find in Wojtyla verges on subjectivism, as if all choices of patients were beyond the reach of any criticism on the part of the doctor. But I do not hold this; I am far from thinking that whatever the patient chooses is right and is right because he chooses it. To hasten to the bedside of the dying parent may be the only right thing to do; a person may later on recognize this objective rightness by reproaching himself bitterly for staying at home, avoiding the risk, and missing the death of the parent. I only say that the doctor has no particular competence for discerning this objective rightness; his medical competence as such does not include any of that personal wisdom that a person needs in deliberating about what risks he personally should take in pursuit of what goods; only the person facing the risks and desiring the goods is competent here.

Of course, there are certain patient choices that the physician is right to protest against, such as the choice to be treated paternalistically. Though many secular bioethicists think that one reasonably exercises one's personal self-possession when one waives one's right to participate responsibly in one's medical care,[10] Wojtyla would certainly think otherwise. For him such a waiver would be not a meaningful exercise of personal self-possession but rather an irresponsible escape from it. He would admonish physicians to avoid any complicity in such escapism. But one could hardly claim to find any offensive paternalism here, as if physicians were lording it over their patients by encouraging them to be themselves as persons and not to relinquish their birthright as persons! And there are other kinds of patient choices that a physician must not uncritically acquiesce in, such as the choice of a woman to keep on smoking heavily during pregnancy. In such a case the patient cannot claim that the choice is too personal for the physician to say anything in criticism of it. She would indeed be speaking like a subjectivist if she were to repel all criticism by saying that her choice is of such a personal nature as to be beyond the reach of any evaluation on the part of the physician. No, there is a higher moral truth, not of the patient's making, not to her liking, an objective moral truth, to which the physician appeals in urging her to stop smoking; but he makes this appeal without in any way asserting himself paternalistically. As long as he gives her all the reasons why smoking during pregnancy is harmful for the child, persuading her rather than intimidating her, he takes her fully seriously as person. And there are still other patient choices that the physician must resist, such as the request for an abortion; the physician who understands abortion for the crime that it is will absolutely refuse any complicity in such a request, and his refusal will have nothing to do with paternalism. And so we have abundant reason for saying that the Wojtylean critique of paternalism that we are developing has nothing to do with subjectivism; the respect that physicians should have for their patients as the ones who in the end integrate their medical diagnosis into their overall existence, has nothing to do with subjectivistically making patients "the measure of all things," as the grandfather of all subjectivists, the ancient sophist Protagoras, famously said in declaring his subjectivism.

I hope it will not seem oversubtle if I offer a further clarification on the respect for the patient as person that underlies the anti-paternalism that Wojtyla would hold. I just said that the physician must stand back when it comes to integrating a medical diagnosis into the overall existence of the patient and must leave this work of integration to the patient. Now it may seem that the basis for having to stand back is the limited knowledge that the physician has of the basic commitments and values of the patient; in experiencing himself from within himself the patient surely knows about these commitments and values with an intimacy and thoroughness that is not possible to the physician and perhaps not even possible to anyone other than the patient himself. Surely the one who has made commitments knows them from within himself like no one else can know them. But it seems important to assume for a moment, for the sake of argument and contrary to fact, that the physician does know the overall existence of the patient just as well as the patient himself knows it, or even better than the patient knows it. And it is very important to see that on this assumption it remains the case that the physician should leave the work of integration to the patient, and that he is guilty of paternalism if he tries to do it for the patient. For each person has a "natural authority" in relation to himself or herself; it is an authority that cannot be delegated, or vested in a proxy, as long as the person is competent. It seems to go with the belonging of persons to themselves of which we spoke above; persons belong inalienably to themselves. Physicians do wrong to their patients as persons if they try to determine for them the place of health among the many other goods in which patients participate, and they would do this wrong even if they were to know better than their patients what is ultimately best for them.

I have spoken of a certain paternalism in the setting of health care as something that is foreign to Wojtyla's personalism. It is quite remarkable – it comes as a great surprise to learn it – that Wojtyla protests against paternalism *even in the setting of our relation to God*. Here is an example of what I mean, which occurs in the course of Wojtyla discussing the anguish people feel when their suffering seems to lack meaning. "Man can put this question [of the meaning of suffering] to God with all the emotion of his heart and with his mind full of dismay and anxiety; and God expects the question and listens to it..."(John Paul II, 1984, p. 13). John Paul thinks that it is by no means impious to cry out in the face of some great suffering, "But why, O Lord, why?" In his book, *Crossing the Threshold of Hope*, he expresses audaciously the religious legitimacy of this cry. He asks how God can justify Himself "before human history, so full of suffering," and he says: "Obviously, one response could be that God does not need to justify Himself to man. It is enough that He is omnipotent. From this perspective everything He does or allows must be accepted." And from this perspective it is indeed impious to press the question, why? But John Paul is not satisfied with this answer; even though it seems to express so profound a reverence for the divine majesty, it is for John Paul paternalistic in a negative sense, or, as we could also express the matter, it tends to make man too much an object of divine providence. He proceeds to give the answer that does satisfy him: "But God, who besides being Omnipotence is Wisdom and – to repeat once again – Love, *desires to justify Himself to mankind* [my italics]. He is not the Absolute that remains outside of the world, indifferent to human suffering. He is Emmanuel, God-with-us..."(John Paul II, 1994, p. 62). Bold words indeed, born of an intimate filial relationship with God: God not only tolerates our question about the meaning of suffering, but wants us to ask it, wants to answer it, and wants us to understand His answer. He takes us so seriously as persons that He wants to "justify Himself" to us. God dwells so intimately with us as to overthrow certain paternalistic models of religion.

One sees the parallel with issues of health care. Physicians must not deliver their judgments about our health care from on high, as if it did not matter whether their patients understand their proposals and the alternatives to them and can integrate them into their larger, more-than-medical lives, nor should physicians act as if their patients had no right to participate in deliberating about those proposals from the point of view of their existence as a whole. They should instead take their patients so seriously as persons that they would rather be challenged by them than proceed with some procedure that is blindly and passively accepted by them and that remains unintegrated in their existence as a whole.

One can also detect Wojtyla's protest against a paternalistic relation to God in an area of his teaching on conjugal morality. When he discusses the rights and wrongs of spacing out children by means of periodic abstinence, he makes it clear that he does not regard the recourse to periodic abstinence as giving evidence of a lack of faith on the part of the spouses. On the contrary, God wants spouses to weigh financial, social, emotional, and other factors in their deliberations about the growth of their family; this is for John Paul an essential part of exercising responsible parenthood. We could perhaps put it like this: God wants spouses to assert their spousal and familial subjectivity and to integrate the good of a new child into the whole of their family life. Some authors have warned against what they call "providentialism," which means the mentality of spouses who live abandoned to divine providence in such a way that they disdain any considerations about spacing out the births of their children. John Paul, too, keeps his distance to providentialism, which for him tends to make spouses too passive in relation to divine providence, and to make them underestimate just how much God wants them to be active collaborators with Him in His providence and not just passive beneficiaries of it. Providentialism can derive in part from the paternalism that is contested by the principles of Wojtyla's personalism.

One could draw out the distinction, implicit in the previous pages, between being an object of divine providence and being a co-subject with God Himself of His providence. The distinction does not originate in Wojtyla's personalism, it is found already in St. Thomas Aquinas. In discussing the natural moral law he says that non-rational creatures are subject to divine providence in one way, and rational creatures in another. Whereas the former are provided for by receiving from God natural inclinations that direct them to their proper acts and ends, the latter are real participants in divine providence, for by understanding the natural law they are capable of providing for themselves and for others.[11] It is only natural to appropriate Wojtyla's language of subject and object and to interpret St. Thomas as distinguishing in effect between being a co-subject with God in exercising His providence and being an object of His providence. What concerns us above all here is the parallel between this distinction and the one between being a co-subject of one's medical treatment and being the object of it.

One sees how Wojtyla's protest against paternalism differs from the affirmation of autonomy on the part of many contemporary bioethicists. For many of them the very idea of being subject to divine providence is offensively paternalistic; for Wojtyla human persons can be subject to divine providence in a way that avoids all paternalism. In other words, Wojtyla acknowledges a creaturely autonomy that is foreign to many of his contemporaries.

In concluding this reflection on paternalism in health care, considered from the point of view of Wojtyla's personalism, we should add that he would also take care to avoid the opposite of paternalism. He would reject the idea that the patient is the only one responsible for his health care and that the physician has only the task of carrying out the wishes of the patient. He rejects it because it tends to degrade the physician to an instrumental means in the hands of the patient. Wojtyla would say that the physician too is a person with his own subjectivity. Just as the personhood of patients is offended by paternalistic treatment, so the personhood of physicians would be offended by reducing them to instruments for the production of the medical benefits chosen by their patients. We have already seen that the physician is often right to resist the wishes of the patient and that he can offer this resistance without any least suspicion of paternalism. We saw among other things that the physician no less than the patient has a conscience and will sometimes have to refuse to comply with patient requests that would involve him, the physician, in some complicity in wrongdoing; in this way, as in other ways, he remains intact as his own person but without in any way asserting himself paternalistically.

5. RESPECTING THE EMBODIED PERSONHOOD OF HUMAN SUFFERERS

We saw above that Wojtyla in his personalism calls particular attention to the interiority and subjectivity of persons. He says that patients are in a particular way endangered in their subjectivity and that physicians have the task of helping them to stay intact as the subjects of their suffering. Now once one focuses on the subjectivity of persons, one is naturally led to the objectivity of their bodily being, which is, as it were, the dimension of their being complementary to their subjectivity. Through my bodily being I am drawn out of my subjectivity and made to exist in a public space where I am available to being perceived by others as their object. In my subjectivity I am hidden from others, but through my bodily objectivity I am exposed to their objectifying inspection, or for that matter to my own objectifying self-inspection.

We can also express the contrast in terms of Wojtyla's *person* and *nature*, and say that thanks to my subjectivity I act through myself and thanks to my objective bodily being I am vulnerable to being acted upon. Now at this point there arises a classic temptation for any personalism: the temptation to exaggerate the contrast between the subjective and the objective in the makeup of human persons and to develop a dualism of interior person and exterior body. Here is an author who fell into this temptation (notice how he uses Buber's terms, I-Thou and I-It, to express the unfortunate dualism):

> Physical nature – the body and its members, our organs and their functions – all of these are a part of "what is over against us," and if we live by the rules and conditions set in physiology or another *it,* we are not *thou....* Freedom, knowledge, choice, responsibility – all these things of personal or moral stature are in us, not *out there*. Physical nature is what is over against us, out there. It represents the world of *its* (Fletcher, 1960, p. 211).

But if I am most of all person in my subjectivity, and if my body is over against my subjectivity, existing as my object, then my body seems almost to fall outside of my personal being, which for its part is almost identified with my non-bodily interiority. Continuing along this line one arrives at the position that the embodiment of persons is extrinsic to them, and so one ends with a kind of spiritualistic personalism.

This spiritualism causes no end of mischief in sexual ethics, including some issues of bioethics. It leads to the mentality of the American feminist who, in the course of debunking all of Christian sexual morality, said, "God doesn't care what we do with each other's bodies, He only cares whether we treat each other as persons." On this view, any and every kind of sexual behavior – fornication, homosexual sex, contraceptive sex – can be made morally acceptable by the presence of an inner attitude of respect for persons. If one says, for example, that homosexual sex is intrinsically disordered, one is told that such a condemnation makes too much of the body and of the purely physiological fact that the sexual organs of man and woman "fit" each other; one is told that it is a far more personalist conception of sexuality to forget about bodily fits and to turn in to the interiority of persons and to look for the attitude of respect for another person. If only this attitude is found, then even the most unconventional sexual behavior can be made morally sound. This attitude is the only morally relevant factor, the exact description of the sexual contact between persons being morally irrelevant. And indeed, given the dualism of person and body, it is hard to resist such an ethical analysis. In fact, I dare to say that not a single article of Christian sexual morality can be made intelligible on the basis of that dualism.

But the question before us is this, what mischief does the dualism cause in our understanding of the physician-patient relation? The answer seems to be that a dualistically minded physician is liable to treat the body of his patient like a broken machine, and that a dualistically minded patient will want to have his body treated like a broken machine. Both of them will see the body of the patient as the property of the patient, and so see it as an object; neither of them will see it as the very patient himself or herself, or as an extension of the person of the patient. They do not think of the subjectivity of the patient as being embodied but as being hidden by the body of the patient. Instead of having a bodily dimension the subjectivity of the patient is taken to be purely spiritual. The physician will think that his treatment of the bodily sickness of his patient is not exactly a treatment of the person who is sick; thus he will feel at liberty to objectify the body of his patient just like he might objectify anything else in nature that he manipulates and controls. And so the patient will find it almost impossible to feel himself to be the subject of his illness, especially when he concurs in this estrangement of himself from his body.

The psychologist Rollo May has perceptively detected some expressions of this estrangement from the body in sick people:

> As a result of several centuries of suppressing the body into an inanimate machine, subordinated to the purposes of modern industrialism, people are proud of paying no attention to the body. They treat it as an object of manipulation, as though it were a truck to be driven till it runs out of gas.
>
> Many disturbances of bodily function, beginning in such simple things as incorrect walking or faulty posture or breathing, are due to the fact that people have all their lives walked, to take only one simple illustration, as though they were machines, and have never experienced any of the feelings in their feet or legs or rest of the body.
>
> The impersonal, separated attitude toward the body is shown also in the way most people, once they become physically ill, react to the sickness. They speak in the passive voice – "I *got* sick," picturing their body as an object just as they would say "I *got* hit by a car." Then they shrug their shoulders and regard their responsibility fulfilled if they go to bed and place themselves completely in the hands of the doctor and the new medical miracle drugs. Thus they use scientific progress as a rationalization for passivity: they know how germs or virus or allergies attack the body, and they also know how penicillin or sulfa or some other drug cures them. The attitude toward disease is not that of the self-aware person who experiences his body as part of himself, but of the compartmentalized person who might express his passive attitude in a sentence like, "The pneumococcus made me sick, but penicillin made me well again" (May, 1953, pp. 107-108).

Now Wojtyla is keenly aware of the dualism underlying this estrangement of persons from their bodies,[12] and he protests against it on many occasions. Perhaps his profoundest reflections on embodiment are found in his "theology of the body,"[13] where he shows that the vocation of human persons to interpersonal love is as it were inscribed in the bodies of man and woman, so that human persons are ordained through their sexually differentiated bodies to the interchange of self-giving love, and are capable of enacting the self-gift through their bodies. Thus the masculine body and the feminine body "participate" intimately in self-giving love, acquiring what Wojtyla calls a "nuptial meaning"; they are intimately incorporated into the subjectivity of man and woman and so cannot possibly be taken as objects existing over against their subjectivity. Thus Wojtyla holds that the sexually differentiated body is not a piece of property held by a person, but is in part the bodily dimension of the person. With this Wojtyla marshals powerful evidence against the person-body dualism and thus contributes indirectly to thinking clearly about what it takes for physicians to treat patients as embodied persons.

But there is an idea in *The Acting Person* that has a more direct bearing on our subject of physician-patient relations. After distinguishing between *person* and *nature*, Wojtyla right away adds that, though they form such a sharp contrast when considered in their essential definitions, they do not give rise to two different subjects of activity in human beings; no, it is one and the same I that now acts through itself, now is acted upon. He resists the temptation to posit one subject for person and one for nature, which would mean that the personal subject could only look over at nature as its object; he acknowledges one human subject in human beings, one who both determines himself in freedom and also undergoes illness.[14] It follows – and this is a consequence of supreme importance for encountering the dualism – that the body is not always only an object in front of me, but *also enters in a way into my subjectivity*, is experienced from within, as only I can experience it. My body is not only an object-body, but a lived body. It is true that neither in this passage nor in his other writings does Wojtyla, as far as I know, explore the phenomenon of the lived body. I would say only that he comes to the threshold of it in this passage, and that we continue along his line by thematizing it.[15]

If as a patient suffering from a certain illness I experience my body only as an object-body, then I cannot remain the subject of the illness; I become depersonalized as patient in the sense that my illness no longer fully belongs to me as person. The treatment undertaken by the physician does not seem to me fully to touch me as person but rather to stop at some point outside of myself. This estrangement of me from my body is reinforced if the physician, too, thinks of my body only as an object-body. But if as a patient suffering from the same illness I experience my body also as lived body, if I know how to dwell in my body as embodied person, experiencing it and my illness from within, as only I can experience it, then I can remain what Wojtyla calls the subject of my illness. The physician can reinforce this subjectivity by always being mindful of dealing with a sick person, always tempering his medical manipulation of my sick body with the consciousness of acting on an embodied person.

It is important to add that a person who knows how to live and experience his embodiment, who knows how to listen to his body, will, when he falls ill, have an experience of his illness that the physician needs to hear about. He will be able to tell the physician many things helpful for reaching a good diagnosis, especially if there is psychosomatic causality at work in the patient. The physician who has freed himself from the dualism of person and body will know how to listen to the reports of the body-sensitive patient; he will expect to learn things directly from the patient that he cannot learn from running tests on the patient.[16]

We conclude, then, that just as Wojtyla's personalism has born much fruit in the study of man and woman and in sexual ethics, and just as it has born much fruit in developing the social teaching of the Church, so it has the potential to bear fruit in understanding the issues of bioethics, included the issue of what it takes for physicians treat their patients as persons.

Franciscan University
Steubenville, Ohio

NOTES

1 I draw here on the discussion of consciousness found in Chapter 1 of Wojtyla's *The Acting Person* (1979). But the term self-presence is my own; I find it more accessible than "the reflexive function of consciousness," which is the way Wojtyla refers to the thing I call self-presence.

2 See above all Wojtyla's seminal paper, "Subjectivity and the Irreducible in the Human Being," in his collection, *Person and Community* (1995).

3 For Wojtyla's thought on self-determination and self-possession I have followed the analyses of *The Acting Person*, Ch. 3.

4 Wojtyla discusses this distinction in *The Acting Person*, Ch. 2, esp. 76-80.

5 See Wojtyla, 1979, pp. 289-290. For example: "Thus conformism consists primarily in an attitude of compliance or resignation, in a specific form of passivity that makes the man-person to be but the subject of *what happens* instead of being the *actor* or *agent* responsible for building his own attitudes and his own commitment in the community. Man then fails to accept his share in constructing the community and allows himself to be carried with and by the anonymous majority" (289).

6 This point is well argued by Germain Grisez, 1997, question 59, "Must a physician tell patients the whole truth about their bad prospects?"

7 The case if different if the anxiety that one can expect in the patient might undermine the patient's competence, temporarily incapacitating the patient for rational judgment; it may then be in order for the physician to invoke the "therapeutic privilege" for withholding certain information about the patient's bad prospects.

8 See Grisez, 1997, p. 275. Grisez's own position seems to agree with the anti-paternalism that I claim to find in Wojtyla's personalism, as when he writes, "patients should not evade their responsibility to make their own decisions; family members should not usurp the patient's role in this matter; and a physician should not be a party to any such evasion or usurpation" (p. 273).

9 It is remarkable that the anti-paternalistic ideal of providing health care that is emerging in this section was already envisioned by Plato, who distinguished between the health care provided by slaves for slaves from the health care provided by free men for free men. "A physician of this kind [the former kind] never gives a servant any account of his complaint, nor asks him for any; he gives him some empirical injunction with an air of finished knowledge, in the brusque fashion of a dictator, and then is off in hot haste to the next ailing servant – that is how he lightens his master's medical labors for him. The free practitioner, who, for the most part, attends free men, treats their disease by going into things thoroughly from the beginning in a scientific way, and takes the patient and his family into his confidence. Thus he learns something from the sufferer, and at the same time instructs the invalid to the best of his power. He does not give his prescriptions until he has won the patient's support, and when he has done so, he steadily aims at producing complete restoration to health by persuading the sufferer into compliance" (Plato, *Laws*, 4.720b-e.)

10 See, for example, Beauchamp and Childress, 2001: "Autonomous choice is a *right*, not a *duty* of patients" (p. 63). See also Engelhardt, 1986, pp. 275-6.

11 See St. Thomas Aquinas, *Summa Theologiae*, I-II, 91, 2. Cf. also I-II, 93, 5.

12 See my paper, "The Estrangement of Persons from Their Bodies," in my collection, *Personalist Papers* (2003). I give here particular attention to Wojtyla's analysis of this estrangement.

13 Wojtyla's "theology of the body" is presented in the series of addresses that he gave as John Paul II between 1979 and 1984 in Rome. They have been collected into one volume: John Paul II, *The Theology of the Body* (1997).

14 See Wojtyla, 1979, pp. 71-85. For example: "For all the sharpness and distinctness...of the differentiation and the contrast [between person and nature], it is impossible to deny that he who acts is simultaneously the one in whom something or other happens" (72).

15 Wojtyla might have thematized it in his discussion (*The Acting Person*, chs. 5 and 6) of what he calls integration. He might have developed there the idea that to fail to live fully one's body is to be deficient with respect to integration.

16 See May, 1953, p. 108: "In overcoming psychosomatic ills or chronic diseases like tuberculosis, it is essential to learn to 'listen to the body' in deciding when to work and when to rest. It is amazing how many hints and guides and intuitions for living comes to the sensitive person who has ears to hear what his body is saying." May refers here to a sensitivity to the lived body the importance of which is not limited to physician-patient relations.

REFERENCES

Beauchamp, T. and J. Childress (2001). *Principles of Biomedical Ethics*, 5th edition. Oxford: Oxford University Press.

Crosby, J. (2003). 'The Estrangement of Persons from Their Bodies,' in J. Crosby, *Personalist Papers*. Washington DC: Catholic University of America Press.

Engelhardt, H.T. (1986). *The Foundations of Bioethics*. New York: Oxford University Press.

Fletcher, J. (1960). *Morals and Medicine*. Boston: Beacon Press.
Frossard, A. (1985). *Be Not Afraid*. New York: St. Martin's Press.
Grisez, G. (1997). *Difficult Moral Questions*. Quincy IL: Franciscan Press.
Guardini, R. (1965). *The Person and the World*. Chicago: Regnery.
John Paul II (1984). *Apostolic Letter on the Christian Meaning of Human Suffering (Salvifici doloris)*. Boston: St. Paul's Editions.
John Paul II (1994). *Crossing the Threshold of Hope*. New York: Alfred A. Knopf.
John Paul II (1997). *The Theology of the Body*. Boston: Pauline Books and Media.
Rollo May, *Man's Search for Himself* (New York: Dell, 1953
Wojtyla, K. (1979). *The Acting Person*. Dordrecht: D. Reidel Publishing Company.
Wojtyla, K. (1981). *Love and Responsibility*. New York: Farrar Straus & Giroux Inc..
Wojtyla, K. (1995). 'Subjectivity and the Irreducible in the Human Being.' in T. Sandok (Trans.), *Person and Community*. New York: Peter Lang.

CHAPTER ELEVEN

JOEL JAMES SHUMAN

JOHN PAUL II AND THE GOODS OF THE BODY (OF CHRIST)

In healing the scattered members come together.
Wendell Berry

... when all is said and done, the law of God is always the one true Good of man.
Pope John Paul II (1993)

1. INTRODUCTION

It has become commonplace to observe that one of the hazards of living in the modern world is that the modern world, for all the very real benefits it offers those of us who inhabit it, divides and scatters us. The circumstances of modernity disconnect us, not simply from the past, from the earth, and from each other, but also from ourselves. The philosopher Alasdair MacIntyre, who is certainly one of modernity's most insightful critics, has referred to this series of disconnections as fragmentation, and has gone so far as to say that fragmentation is a predominant characteristic of life in modern societies. As the breadth and depth of human knowledge and technological capability have increased, enquiry has necessarily become more specialized. The great success of the physical and natural sciences have made them the paradigm for all human knowing, and the so-called "social sciences" have imitated their methods and aspired to equal their public status.

At the same time, an increased emphasis upon and appreciation for the individual and her rights has been paralleled by the diminishment of the guiding influence of extended families, local communities and religious traditions. Instead of looking to these traditional sources, we have become increasingly dependent upon a large and growing cadre of experts to help us negotiate our worlds. We rely upon these women and men to tell us, typically through the machinations of various bureaucratic apparatuses, about the worlds we inhabit. By implication, we depend on them also to tell us how we should live our lives, at least with respect to this or that development in their respective fields of expertise. Yet because enquiry has become so specialized, the province of these various experts is theirs alone, and they tend to have little if anything to do with or say to each other. Just so, we are increasingly unable to say or even to imagine what it might mean to live well over time, to live along a consistent trajectory corresponding to a fulsome vision of a

C. Tollefsen (Ed.), John Paul II's Contribution to Catholic Bioethics, pp. 169–180.

good life. We know a great deal about what we are, but very little about what we are for, and so we are left instead to cobble our lives together as best we can, based upon the fragments of knowledge we pick up here and there. We have learned to call this individualized cobbling together "freedom" (MacIntyre, 1982, pp. 36-108; Berry, 1972, pp. 86-168).

The fragmentary character of modern life is displayed as much as anywhere in the array of questions arising from our interactions with medicine. Along with the very significant growth of medicine's capabilities and the concomitant expansion of its status within contemporary culture has come a growing confusion about its proper utilization. Medicine seems ceaselessly to amaze us with its discovery of new ways to police, heal and improve our bodies. Yet, as many have observed, such discoveries frequently outpace our confidence about whether or how we ought to use them. For example, the simple possession of the capacity to predict or diagnose life-threatening conditions in embryos and fetuses, or to sustain the biological life of the otherwise dying by so-called "artificial" or "extraordinary" means is not in and of itself an indication that such capacities are always or even sometimes justifiably acted upon. Here too, we require the help of experts.

The emergence and maturity of the discipline of bioethics is frequently seen as a necessary response to these difficult matters. Bioethicists are increasingly seen as the people who possess the expert knowledge to help us make good decisions about the use of medicine and especially of complex medical technology. Yet, in the very growth and popular acceptance of bioethics as a discipline, we witness a furthering of the fragmentation that characterizes our lives. As Carl Elliot has recently written, the emergence of the expert bioethicist, while perhaps helping some of us answer some questions, also produces a danger:

> It is the danger of people losing trust in their own moral judgments until they have been confirmed by an expert. And not only this: it is losing confidence in one's moral experience as well. I have not identified the moral features of this case correctly until the ethicist has confirmed it; I have not told the story correctly until the ethicist says I have; this is not an ethics case (as opposed to a social work case, or a psychiatric case, or a case of institutional politics) until the ethicist says it is. Ethics is something done by Them. My own experience needs to be validated before it counts. It needs their stamp of approval (Elliot, 2002, p. 15).

What Elliot does not discuss is that the kinds of moral knowledge and experience he is concerned we are losing require for their development and sustenance participation in a moral tradition. And it is precisely such moral traditions that are threatened by the fragmenting tendencies of the modern world. Christians – especially Protestants – are no less victims of this fragmentation and subsequent reliance on experts than anyone else is. We have learned over time to treat our faith as an idea. Our spirituality is something we choose individually and hold privately and interiorly, for our own personal benefit, rather than as the basis for an entire way of life sustained by the practices of a community. Just to the extent we have become Gnostics who accept the modern relegation of "religion" to the largely irrelevant sphere of private, experiential interiority, we have come either to forget or ignore that Christianity is much more than this (Lee, 1987). It is an account of God and the people of God, one that tells us, often in no uncertain terms, how we should live. Christianity tells us, in other words, what we are for. It is an account, as the subject of this volume has said, of our one true Good.

Another way of saying this is that the Christian narrative is a story about God's steadfast love for and restoration and fulfillment of the material creation, which God called and continues to call "good." And because we ourselves are members of the material creation, the Christian narrative is a story about the restoration of our bodies and their goods. Just so, the Christian story is a story about health and healing. A significant part of what Christians call salvation is our being made part of a people whose common life teaches us properly to value the gift that is our body. Christianity is as much about eating and drinking and working and playing and making love; as much about facing sickness and death and caring for the sick and dying as it is about prayer and contemplation. In a letter to a close friend, written a few short months before her death from an incurable and at that time largely untreatable disease, Flannery O'Connor offered a salient reminder of this:

> For me it is the virgin birth, the Incarnation, the resurrection that are the true laws of the flesh and the physical. Death decay, destruction are the suspension of these laws. I am always astonished at the emphasis the Church puts on the body. It is not the soul she says will rise but the body, glorified. I have always thought that purity was the most mysterious of the virtues, but it occurs to me that it would never have entered the human consciousness to conceive of purity if we were not to look forward to a resurrection of the body, which will be flesh and spirit united in peace, in the way they were in Christ. The resurrection of Christ seems the high point of the law of nature (O'Connor, 1979, p. 100).

To call the resurrection of Christ "the high point of the law of nature" seems terribly anachronistic in a world where each day brings news that medical science has unlocked another of the secrets of the human body. Yet O'Connor reminds Christians of a truth that we cannot afford to forget; our bodies are, when all is said and done, God's good creatures. As such, our bodies are indispensable to the achievement of our *telos* within the created order. God has created us to be friends – of God, of one another and of the rest of the creation. Such friendships require bodily presence over time, a presence that is learned and cultivated in the gathered community of the people of God. This must be the starting place for any explicitly Christian bioethics. And it is precisely here that all Christians, and especially Protestants, can learn from the work of Pope John Paul II. The greatest contribution of the pontiff to bioethics is this: he reminds us that there is no more truthful description of what our bodies are, and so of what they are for, than the theological description we learn from being gathered as members of the Body of Christ. Although he sometimes leaves it to others adequately to draw his arguments to sufficiently thick conclusion, John Paul II must nonetheless be regarded as having made a significant contribution toward the recovery of an explicitly theological bioethics.

2. FAITH AND THE MORAL LIFE

The overarching impression one takes from reading the work of John Paul II is of a moral theology in which human action may properly be understood and described only as part of a total theological ecology, in which divine and human action are linked without the possibility of extrication. "The moral life," he explains, "presents itself as the response due to the many gratuitous initiatives taken by God out of love for man" (John Paul II, 1993, no. 10).[1] Women and men act properly and fully only as players in a drama enacted in the theatre of the prior and simultaneous saving action of God. This action is made visible in history in the stories of God's election of Israel, Jesus and the Church, which in turn form the backdrop for all truly human action.

> The good is belonging to God, obeying him, walking humbly with him in doing justice and loving kindness (cf. Mic 6:8). Acknowledging the Lord as God is the very core, the heart of the Law, from which the particular precepts flow and toward which they are ordered. In the morality of the commandments the fact that the people of Israel belong to the Lord is made evident, because God alone is the One who is good. Such is the witness of Sacred Scripture, imbued in every one of its pages with a lively perception of God's absolute holiness: "Holy, holy, holy is the Lord of hosts" (Is 6:3) (1993, no. 11).

At the center of God's saving acts, and so at the center of all history, is the Gospel embodied in the life of Jesus of Nazareth. (John Paul II, 1979, no. 1) Jesus is the Word made flesh, the incarnation of the creating, redeeming and sustaining God. And just because the creating, redeeming, sustaining God has become human, humanity is given in that single human life a window into true "Human nature." It is "only in the mystery of the Incarnate Word," as the Vatican II document *Gaudium et Spes* says, that "the mystery of man take[s] on light." Jesus is the exemplary human and the possibility of our being truly human:

> And the Council continues: "He who is the 'image of the invisible God' (Col 1:15), is himself the perfect man who has restored in the children of Adam that likeness to God which has been disfigured ever since the first sin. Human nature, by the very fact that it was assumed, not absorbed in him, has been raised in us also to a dignity beyond compare. For, by his Incarnation, he, the son of God, *in a certain way united himself with each man*. He worked with human hands, he thought with a human mind. He acted with a human will, and with a human heart he loved. Born of the Virgin Mary, he has truly been made one of us, like to us in all things except sin", he, the redeemer of man (1979, no. 8, italics original).

That the agent and means of human redemption is at once the perfect, exemplary human and the perfect embodiment of God's steadfast love for humankind suggests that the salvation effected by Jesus always and everywhere involves the witness of those who by baptism are made part of his life and teaching.[2] Faith and morality – that is, a particular, concrete way of life formed and disciplined by the common life of a community – are therefore inextricably connected.[3] Just as Jesus displays to the world God's ultimate good intention for the creation, those who have been made members of his body through baptism are called to display that intention by imitating his life and obeying his teachings (1993, no. 21). "The way and at the same time the content of [human] perfection consists in the following of Jesus, *sequelae Christi*" (1993, no. 19). This is how women and men learn what they are and how they are to live; not from the highly individualized, narcissistic, consumer culture of our time,[4] but from the Gospel of Jesus of Nazareth:

> The man who wishes to understand himself thoroughly – and not just in accordance with immediate, partial, often superficial, and even illusory standards and measures of his being – he must with his unrest, uncertainty and even his weakness and sinfulness, with his life and death, draw near to Christ. He must, so to speak, enter into him with all his own self, he must "appropriate" and assimilate the whole of the reality of the Incarnation and Redemption in order to find himself (1979, no. 10).

Insofar as human being and action are properly specified and judged according to the self-giving of God in Christ, the typical distinction between freedom and obedience – or, to use a more explicitly theological idiom, that between gospel and law – is at best problematic. Women and men are properly free, but freedom is not reducible to the autonomous exercise of an individual human will. Rather, true freedom is ultimately determined by the objective moral order established by God in creation and restored by the Gospel (1993, nos. 32-34). True autonomy is the free participation of the person in this order:

> Man's *genuine moral autonomy* in no way means the rejection but rather the acceptance of the moral law, of God's command: "The Lord God gave this command to the man..." (Gen 2:16). *Human freedom and God's law meet and are called to intersect*, in the sense of man's free obedience to God and of God's completely gratuitous benevolence toward man. Hence obedience to God is not, as some would believe, *heteronomy*, as if the moral life were subject to something all-powerful, absolute, extraneous to man and intolerant of his freedom. If in fact a heteronomy of morality were to mean a denial of man's self-determination or the imposition of norms unrelated to his good, this would be in contradiction to the revelation of the Covenant and of the redemptive Incarnation. Such a heteronomy would be nothing but a form of alienation, contrary to divine wisdom and to the dignity of the human person (1993, no. 41).

The foundation of these claims is the fundamental Christian conviction that God values, and so cares actively for the creation. That God has established a moral order in creation, restored it through God's saving acts, and called and enabled women and men to participate in it are signs of this care. This in turn requires us to rethink what we mean when we say "bioethics." Within this scheme of God's salvific activity, bioethics is the name we may give to that discourse that helps us determine in particular cases how to live well as bodied creatures whose lives are forever dependent upon God, one another, and the earth. John Paul II wishes us to understand that the proper basis of such a discourse is the Christian gospel, the good news that God in Christ is restoring our world and all its members to their original created goodness.

3. "CALLED TO BE GOOD BODIES"

Another way of putting this is to say that the gospel teaches us to be good bodies. This is one of the most significant of the senses in which it is a "gospel of life"; life is the most basic – though not the ultimate – of human goods. Biological life alone is good, but only provisionally, for in its fullest sense human life "consists in communion with the Father, to which every person is freely called in the Son by the power of the Sanctifying Spirit. It is precisely in this 'life' that all the aspects and stages of human life achieve their full significance" (John Paul II, 1995, no. 1). Biological life is thus a gift from God that is in some sense, to coin a phrase, the material condition of possibility for human life in its fullness. "In giving life to man, God demands that he love, respect and promote life. Life thus becomes a commandment, and the commandment is itself a gift" (1995, no. 52). True life, and so the true value of life, are given in the gospel. "Through the words, the actions and the very person of Jesus, man is given the possibility of 'knowing' the complete truth concerning the value of human life. From this 'source' he receives, in particular, the capacity to 'accomplish' this truth perfectly (cf. Jn 3:21), that is, to accept and fulfill completely the responsibility of loving and serving, of defending and promoting human life" (1995, no. 29).

Just so, we are led to consider again the significance of the incarnation, perhaps the ultimate affirmation of the significance – and the goodness – of human life. The incarnation reminds us that the body is an irreducible aspect of human life. The body is not a passive machine that is merely an instrument of the human person. Rather, it is fundamental to our participation in the moral order established in creation and restored by God's saving activity, at once the subject and the object of our moral lives.

It is therefore inappropriate to suggest that the free exercise of human rationality licenses us to alter the circumstances of our existence in whatever ways we see fit (1993, nos 48-49). There are clearly better and worse ways of being a human body. Insofar as I am my body, the way I live is either consistent or not with the moral order established in creation and restored in the life, death and resurrection of Jesus. Among the central tenets of this order is the notion that humans have what Gilbert Meilander has called a "natural history" which should broadly be respected (Meilander, 1995, pp. 37-59). We are conceived, birthed and nurtured by our parents and our communities; we grow and mature, and in time become parents and nurturers ourselves. Eventually, we grow old; our capacities gradually diminish, and we become once again dependent upon the nurture of others as we approach the ends of our natural lives. And although our respective individual histories differ greatly, this trajectory from conception to death is in some sense our destiny before God, one that is being restored by God's gracious presence to the community of God's people. This means that:

> The Church cannot abandon man, for his "destiny", that is to say his election, calling, birth and death, salvation and perdition, is so closely and unbreakably linked with Christ. We are speaking precisely of each man on this planet, this earth that the Creator gave to the first man, saying to the man and the woman: "subdue it and have dominion." Each man in all the unrepeatable reality of what he is and what he does, of his intellect, and will, of his conscience and heart. Man who in his reality has, because he is a "person", a history of his life that is his own and, most important, a history of his soul that is his own. Man who, in keeping with the openness of his spirit within and also with the many diverse needs of his body and his existence in time, writes this personal history of his through numerous bonds, contacts, situations and social structures linking him with other men, beginning to do so from the first moment of his existence on earth, from the moment of his conception and birth (1979, no. 14).

Of course, we do not all experience our personal histories in precisely the same way. Although God in Christ is restoring the creation, that restoration remains unrealized. Illness and disability of various kinds frequently disrupt our lives, and we are individually and collectively often frustrated by the limits placed on us by our finitude. As rational creatures that have at our disposal a wealth of knowledge and a great deal of sophisticated biotechnology, we have the capacity to intervene and manipulate the circumstances of our bodily existence. We can in other words change certain aspects of the course of our lives and alter our personal histories. Indeed, this ability displays in a profound way what it means to be human; women and men are given by God both the capacity and the mandate to work to improve certain of the circumstances of our lives. "Called to be fruitful and multiply, to subdue the earth and to exercise dominion over the lesser creatures (cf. Gen 1:28), man is ruler and lord not over things but especially over himself, and in a certain sense over the life which he has received and which he is able to transmit through procreation, carried out with love and respect for God's plan" (1995, no. 52).

Yet the ability to manipulate our existence also has the potential to evidence the depths of our brokenness – our alienation from God, from each other, and from the earth. For the very freedom and rationality that are the basis of our being human also afford us the ability to be less than human, and so to increase the magnitude of our alienation:

> This is why all phases of present-day progress must be followed attentively....What is in question is the advancement of persons, not just the multiplying of things that people can use. It is a matter, as a contemporary philosopher has said and as the Council has stated – not so much of "having more" as of "being more." Indeed there is already a real perceptible danger that, while man's dominion over the world of things is making enormous advances, he should lose the essential threads of dominion and in various ways let his humanity be subjected to the world and become himself something subject to manipulation in many ways – even if the manipulation is often not perceptible directly – through the whole of the organization of community life, through the production system and through pressure from the means of social communication. Man cannot relinquish himself or the place in the visible world that belongs to him; he cannot become the slave of things, the slave of economic systems, the slave of production, the slave of his own products (1979, no. 16).

The Christian gospel reminds us that our capacities to use technology to change the circumstances of our lives are limited and measured goods that in fact have the capacity to enslave and perhaps even destroy us. This is one of the more prominent and persistent themes in the critique of modernity characterizing so much of John Paul II's work. It is not at all self-evident that every technological development represents genuine "progress," and the term must be applied with care in particular cases:

> Does not the previously unknown immense progress – which has taken place especially in the course of this century – in the field of man's dominion over the world itself reveal – to a previously unknown degree – that manifold subjection "to futility"? It is enough to recall certain phenomena, such as the threat of pollution of the natural environment in areas of rapid industrialization, or the armed conflicts continually breaking out over and over again, or the prospectives of self-destruction through the use of atomic, hydrogen, neutron and similar weapons, or the lack of respect for the life of the unborn. The world of the new age, the world of space flights, the world of previously unattained conquests of science and technology – is it not also the world of "groaning in travail" that "waits with eager longing for the revealing of the sons of God" (1979, no. 8)?

Indeed, in a world where the language of will and right has largely replaced the language of ends, it is not at all clear that the notion "progress" is even intelligible. To speak of progress presumes movement along a trajectory toward an end, yet modern thought largely precludes speaking about ends. In the Christian tradition, the proper end of human life is friendship with God – a goal quite different in kind from those achieved through technological development. While it is true that some technological achievements can help facilitate the attainment of certain of the proximate ends requisite to a good human life, it is not at all clear that this is necessarily the case. The most significant criterion for assessing technological achievement is whether it threatens our being friends of God:

> This question must be put by Christians, precisely because Jesus Christ has made them so universally sensitive about the problem of man. The same question must be asked by all men, especially those belonging to the social groups that are dedicating themselves actively to development and progress today. As we observe and take part in these

> processes we cannot let ourselves be taken over merely by euphoria or be carried away by one-sided enthusiasm for our conquests, but we must all ask ourselves, with absolute honesty, objectivity and a sense of moral responsibility, the essential questions concerning man's situation today and in the future. Do all the conquests attained until now and those projected for the future for technology accord with man's moral and spiritual progress? In this context is man, as man, developing and progressing or is he regressing and being degraded in his humanity" In men and "in man's world", which in itself is a world of moral good and evil, does good prevail over evil (1979, no. 15)?

4. THE PROBLEM OF SUFFERING

Yet, in spite of admonitions such as these, the temptation for us to depend without reserve upon technology remains a significant one. This is primarily because technology has proven such an effective weapon in the struggle against what we understand as one of the few remaining common enemies of humanity, suffering. Modern thought has no way to make sense of the human experience of suffering. Just so, it is in our conflict with, and our frequently legitimate desire to escape various kinds of suffering that the temptation to use technology to become less than human is most pronounced. Here the basic Christian conviction that in this era God rules the world through the sufferings of Christ and his body are at once seminally important and profoundly tested.

This is of course among the most difficult and complex matters with which Christianity has struggled. There exists from the beginning of the Christian conversation a clear, strong, and some would say dominant voice claiming that although suffering itself is a consequence of sin and so not a good, God sometimes uses suffering as a modality of God's redemptive work. It is therefore consistent with the idea of real human progress – what the Pope calls "man's moral and spiritual progress" – that some kinds of suffering are possibly to be endured, rather than escaped. Yet because suffering is frequently so awful, because it "is always a trial – at times a very hard one – to which humanity is subjected" (John Paul II, 1984, no. 23), it is difficult to permit ourselves unnecessarily to endure any suffering – especially in the midst of our current overwhelming optimism that our coming to possess the means to overcome all suffering is only a matter of our investing adequate time and resources. Increasingly, we are persuaded not only that suffering is undesirable, but also that it is eradicable and should not be endured.

The work of John Paul II suggests that this attitude reflects an attenuated understanding, both of the experience of suffering and of the place of suffering in the human condition. Suffering is not limited to the experience of physical hardship, for the "field of human suffering is much wider, more varied, and multidimensional. Man suffers in different ways, ways not always considered by medicine, not even in its most advanced specializations" (John Paul II, 1984, no. 5). To be sure, sickness is a significant cause of human suffering, and John Paul II never claims that the totality of suffering caused by sickness ought simply to be endured for the sake of some abstract spiritual benefit. The resources afforded by modern technology are themselves goods in the service of human life and flourishing, and to deny their use in the name of learning from the suffering such a denial would cause is absurd. Yet this is not the case absolutely. Indeed, we are especially susceptible to self-deception at this point. Suffering's sheer complexity suggests that at least some of the activities we think of as eradicating suffering are in fact, because they threaten to separate us from God, only exchanging one kind of suffering for another.

> A certain idea of this problem comes to us from the distinction between physical suffering and moral suffering. This distinction is based upon the double dimension of human being and indicates the bodily and spiritual element as the immediate or direct subject of suffering. Insofar as the words "suffering" and "pain" can, up to a certain degree, be used as synonyms, physical suffering is present when the body is hurting in some way, whereas moral suffering is "pain of the soul." In fact, it is a question of pain of a spiritual nature, and not only of the "psychological" dimension of pain which accompanies both moral and physical suffering. The vastness and the many forms of moral suffering are certainly no less in number than the forms of physical suffering. But at the same time, moral suffering seems, as it were, less identified and less reachable by therapy (1984, no. 5).

The experience of suffering is in fact an integral aspect of human being; "what we express by the word... seems to be particularly essential to the nature of man." Although all creatures may experience the unhappy consequences of creation's brokenness, suffering in the fullest sense "manifests in its own way that depth which is proper to man, and in its own way surpasses it. Suffering seems to belong to man's transcendence: it is one of those points in which man is in a certain sense 'destined' to go beyond himself, and he is called to this in a mysterious way" (1984, no. 2). Sufferings direct the attentions of women and men to the fact that they are not fully in control of their lives and are ultimately dependent for their existence on others and perhaps on an Other. In the presence of suffering, a space is created in which it is possible to benefit from and participate in the work of God in the world (John Paul II, 1980, no. 12).

To claim that some suffering may be endured because it might be transformed to serve as a modality of redemption is to risk the trivialization of the horrific. Yet, such a claim is unavoidably part of the Christian Gospel. For if the phenomenon of human suffering affords to those who suffer and those who surround the suffering the possibility of participating in the work of God in the world, then suffering is much more than the sum of the physical and psychic pain it causes. One of the more commonly voiced concerns about suffering is that it is burdensome and disrupts the lives both of the person suffering and those surrounding her (1980, no. 27). But just to the extent that suffering affords a connection to God, it becomes difficult to see it only as disruptive to life. Rather, it may be constitutive of life by reminding us of our fundamental interdependence with one another, and our final dependence upon God.

But this is only the case if God is the One revealed in the biblical story of Israel and in the life, death and resurrection of Jesus of Nazareth. Here we are brought back to the centrality of the cross and its inescapable significance for the ongoing life of the Christian community. "One can say that with the passion of Christ all human suffering has found itself in a new situation" (1984, no. 19). The endurance of suffering and the way the community cares for its suffering members are not existential sleights of hand, but virtues given by the grace of a crucified God, which enable the people of God to be a faithful witness to the cross of Christ, an event that points constantly toward his resurrection, and so toward the eradication of all suffering.

> The foundation of the eschatological fulfillment is already contained in the cross of Christ and in his death. The fact that Christ was "raised the third day" constitutes the final sign of the messianic mission, a sign that perfects the entire revelation of merciful love in a world that is subject to evil. At the same time it constitutes the sign that foretells "a new heaven and a new earth," when God "will wipe away every tear from their eyes, there will be no more death, or mourning, no crying nor pain, for the former things have passed away" (1980, no. 84).

5. THE PROBLEM OF THE POLITICAL

This is a compelling theological vision. Yet, it raises a significant question that to my mind is never satisfactorily addressed by the work of John Paul II. Where is the particular theological politics that can sustain his strong, even radical, Christology and the deep Christocentricity of his moral vision? For surely it is one thing to say "each man, in his suffering, can also become a sharer in the redemptive suffering of Christ" (1984, no. 19), but another thing altogether to name and describe the constellation of human relationships that might make it possible for some to endure certain kinds of suffering in such ways that their lives might become Gospel – that is, "good news" – to the watching world. The Pope says rightly "In the eschatological fulfillment mercy will be revealed as love, while in the temporal phase, in human history, which is at the same time the history of sin and death, love must be revealed above all as mercy and must be actualized as mercy" (1980, no. 85). Yet he writes much more about the significance of enduring suffering in anticipation of the mercy of God, than of specific ways in which Christians can be the mercy of God to the sufferings of others.

Admittedly, he does allude to the significance of the Church as a kind of counterculture whose way of life stands in sharp contrast to that of the world. In his call for "a new culture of human life" in *Evangelium Vitae*, he exhorts the Church to be "a people of life because God, in his unconditional love, has given us the Gospel of life and has by this same Gospel we have been transformed and saved" (1995, no. 78). Certainly this is true; just to the extent that the Christian Gospel is a Gospel of life, those who have had their lives taken up and made part of that Gospel ought to have a special appreciation for the gift that is life. Seldom, however, does John Paul II make clear the kinds of social and economic practices that are requisite to the kinds of appreciative celebrations of life he desires. He speaks frequently and eloquently about Jesus, but says surprisingly little about Jesus' teachings or the explicitly social and political nature of Jesus life and work.[5]

It is altogether fitting, for example, to stand opposed to abortion as a moral evil and to admonish Christians for their acquiescence to contemporary culture's often facile justifications of abortion on demand. And it is equally appropriate to call upon men to take responsibility for their sexual behavior and to admonish various kinds of institutions for their role in the trivialization of life in the womb (1995, nos. 58-62). Similarly praiseworthy is the Pope's condemnation of contemporary culture's easy acceptance of physician-assisted suicide as part of a broad obsession with death and its control (1995, nos. 64-66). Yet these claims all would be strengthened – not to mention rendered more intelligible – by showing explicitly how the christological and ecclesiological convictions of the Christian community give rise to a way of life that includes the free sharing of all manner of resources, from time and presence to material and monetary assets to hospitable space. This kind of an explicit account of the church as a radical *alterna civitas* is largely missing from the Pope's work.

Let us assume, however, that such an account *is* implicit in that work. If this is the case, our attention is called again to the significance of witness, both as a category in moral theology and an irreducible aspect of the life of the church. And witness is in the Christian tradition almost always a function of community. The real goods of the human body are learned, pursued, and achieved as the body of the individual is by baptism made part of the one Body of Christ. There, sickness and suffering are no less onerous than they are any place else. Yet, their burdens are shared by a gathering of persons devoted to being God's presence to the world, a gathering in which "when one member suffers, all suffer together with it." John Paul II is eloquent in his reminding all Christians that sharing in the life, and so in the sufferings of Christ is a goal to which we are to aspire. We need only to be reminded that it is a goal that will only be achieved as we join with one another – and help one another – in its pursuit. This is where our true healing lies, in Christ, as we, the scattered members of his body, come once again together (Berry, 1990, pp. 9-13).

King's College
Wilkes-Barre, Pennsylvania

NOTES

1 John Paul's encyclicals have been compiled in a single volume in Miller, 1996. References in the text are to the encyclicals by date.

2 For an especially compelling account of the significance of witness in the Christian life generally and the work of John Paul II in particular, see Hauerwas, 2001, pp. 205-241.

3 The connection of moral theology to theology proper, specifically Christology and soteriology, is a significant concern for John Paul II. In *Veritatis Splendor*, no. 37, he notes: "certain moral theologians have introduced a sharp distinction, contrary to Catholic doctrine, between an *ethical order*, which would be human in origin and of value for this world alone, and an *order of salvation*, for which only certain intentions and interior attitudes regarding God and neighbor would be significant. This has then led to an actual denial that there exists, in Divine Revelation, a specific and determined moral content, universally valid and permanent.... No one can fail to see that such an interpretation of the autonomy of human reason involves positions incompatible with Catholic teaching."

4 Here see Shuman and Meador, 2002.

5 Here see Yoder, 1972.

REFERENCES

Berry, W. (1972). *A Continuous Harmony*. San Diego: Harcourt & Brace.
Berry, W. (1990). *What Are People For?* New York: North Point Press.
Elliot, C. (2002). 'Miss Lonelyhearts Visits the Hospital' (unpublished paper).
Hauerwas, S. (2001). *With the Grain of the Universe: The Church's Witness and Natural Theology*. Grand Rapids: Brazos.
John Paul II. (1979). *Redemptor Hominis* in Miller, M. (Ed.) (1996). *The Encyclicals of John Paul II*. Huntington, IN: Our Sunday Visitor Press.
John Paul II. (1980). *Dives in Misericordia* in Miller, M. (Ed.) (1996). *The Encyclicals of John Paul II*. Huntington, IN: Our Sunday Visitor Press.
John Paul II. (1984). *Salvifici Doloris* in Miller, M. (Ed.) (1996). *The Encyclicals of John Paul II*. Huntington, IN: Our Sunday Visitor Press.

John Paul II. (1993). *Veritatis Splendor* in Miller, M. (Ed.) (1996). *The Encyclicals of John Paul II.* Huntington, IN: Our Sunday Visitor Press.
John Paul II. (1995). *Evangelium Vitae* in Miller, M. (Ed.) (1996). *The Encyclicals of John Paul II.* Huntington, IN: Our Sunday Visitor Press.
Lee, P. (1987). *Against the Protestant Gnostics*. Oxford: Oxford University Press.
Miller, M. (Ed.) (1996). *The Encyclicals of John Paul II*. Huntington, IN: Our Sunday Visitor Press.
MacIntyre, A. (1982). *After Virtue* (2nd ed.). Notre Dame, IN: University of Notre Dame Press.
Meilander, G. (1995). *Body, Soul and Bioethics*. Notre Dame, IN: University of Notre Dame Press.
Shuman, J., & K. Meador (2002). *Heal Thyself: Spirituality, Medicine and the Distortion of Christianity*. New York: Oxford University Press.
Yoder, J. (1972). *The Politics of Jesus*. Grand Rapids: Eerdmans.

NOTES ON CONTRIBUTORS

Mark J. Cherry, Assistant Professor of Philosophy at St. Edward's University, in Austin Texas, U.S.A.

Gavin T. Colvert, Associate Professor of Philosophy at Assumption College in Worcester, Massachusetts, U.S.A.

John F. Crosby, Professor of Philosophy at the Franciscan University at Steubenville, in Steubenville, Ohio, U.S.A.

Laura L. Garcia, Department of Philosophy, Boston College, Chestnut Hill, Massachusetts, U.S.A.

Luke Gormally, Senior Research Fellow, The Linacre Centre for Healthcare Ethics, London, England and Research Professor, Ave Maria School of Law, Ann Arbor, Michigan, U.S.A.

Patrick Lee, Professor of Philosophy at the Franciscan University at Steubenville, in Steubenville, Ohio, U.S.A.

Andrew Lustig, Director, Program on Biotechnology, Religion, and Ethics, and Research Scholar in Religious Studies, Rice University, Houston, Texas, U.S.A.

William E. May, Michael J. McGivney Professor of Moral Theology at the John Paul II Institute for Studies on Marriage and Family at The Catholic University, Washington, D.C., U.S.A.

Michael J. Murray, Arthur and Katherine Shadek Professor in the Humanities, Franklin and Marshall College, Lancaster, Pennsylvania, U.S.A.

Joel J. Shuman, Assistant Professor, Department of Theology, King's College, Wilkes-Barre, Pennsylvania, U.S.A.

Christopher Tollefsen, Associate Professor of Philosophy, University of South Carolina, Columbia, South Carolina, U.S.A. and Ann and Herbert W. Vaughan Fellow, the James Madison Program, Department of Politics, Princeton University, Princeton, New Jersey, U.S.A.

INDEX

Philosophy and Medicine

1. H. Tristram Engelhardt, Jr. and S.F. Spicker (eds.): *Evaluation and Explanation in the Biomedical Sciences.* 1975 ISBN 90-277-0553-4
2. S.F. Spicker and H. Tristram Engelhardt, Jr. (eds.): *Philosophical Dimensions of the Neuro-Medical Sciences.* 1976 ISBN 90-277-0672-7
3. S.F. Spicker and H. Tristram Engelhardt, Jr. (eds.): *Philosophical Medical Ethics.* Its Nature and Significance. 1977 ISBN 90-277-0772-3
4. H. Tristram Engelhardt, Jr. and S.F. Spicker (eds.): *Mental Health.* Philosophical Perspectives. 1978 ISBN 90-277-0828-2
5. B.A. Brody and H. Tristram Engelhardt, Jr. (eds.): *Mental Illness.* Law and Public Policy. 1980 ISBN 90-277-1057-0
6. H. Tristram Engelhardt, Jr., S.F. Spicker and B. Towers (eds.): *Clinical Judgment.* A Critical Appraisal. 1979 ISBN 90-277-0952-1
7. S.F. Spicker (ed.): *Organism, Medicine, and Metaphysics.* Essays in Honor of Hans Jonas on His 75th Birthday. 1978 ISBN 90-277-0823-1
8. E.E. Shelp (ed.): *Justice and Health Care.* 1981 ISBN 90-277-1207-7; Pb 90-277-1251-4
9. S.F. Spicker, J.M. Healey, Jr. and H. Tristram Engelhardt, Jr. (eds.): *The Law-Medicine Relation.* A Philosophical Exploration. 1981 ISBN 90-277-1217-4
10. W.B. Bondeson, H. Tristram Engelhardt, Jr., S.F. Spicker and J.M. White, Jr. (eds.): *New Knowledge in the Biomedical Sciences.* Some Moral Implications of Its Acquisition, Possession, and Use. 1982 ISBN 90-277-1319-7
11. E.E. Shelp (ed.): *Beneficence and Health Care.* 1982 ISBN 90-277-1377-4
12. G.J. Agich (ed.): *Responsibility in Health Care.* 1982 ISBN 90-277-1417-7
13. W.B. Bondeson, H. Tristram Engelhardt, Jr., S.F. Spicker and D.H. Winship: *Abortion and the Status of the Fetus.* 2nd printing, 1984 ISBN 90-277-1493-2
14. E.E. Shelp (ed.): *The Clinical Encounter.* The Moral Fabric of the Patient-Physician Relationship. 1983 ISBN 90-277-1593-9
15. L. Kopelman and J.C. Moskop (eds.): *Ethics and Mental Retardation.* 1984 ISBN 90-277-1630-7
16. L. Nordenfelt and B.I.B. Lindahl (eds.): *Health, Disease, and Causal Explanations in Medicine.* 1984 ISBN 90-277-1660-9
17. E.E. Shelp (ed.): *Virtue and Medicine.* Explorations in the Character of Medicine. 1985 ISBN 90-277-1808-3
18. P. Carrick: *Medical Ethics in Antiquity.* Philosophical Perspectives on Abortion and Euthanasia. 1985 ISBN 90-277-1825-3; Pb 90-277-1915-2
19. J.C. Moskop and L. Kopelman (eds.): *Ethics and Critical Care Medicine.* 1985 ISBN 90-277-1820-2
20. E.E. Shelp (ed.): *Theology and Bioethics.* Exploring the Foundations and Frontiers. 1985 ISBN 90-277-1857-1

Philosophy and Medicine

21. G.J. Agich and C.E. Begley (eds.): *The Price of Health.* 1986
ISBN 90-277-2285-4
22. E.E. Shelp (ed.): *Sexuality and Medicine.* Vol. I: Conceptual Roots. 1987
ISBN 90-277-2290-0; Pb 90-277-2386-9
23. E.E. Shelp (ed.): *Sexuality and Medicine.* Vol. II: Ethical Viewpoints in Transition. 1987 ISBN 1-55608-013-1; Pb 1-55608-016-6
24. R.C. McMillan, H. Tristram Engelhardt, Jr., and S.F. Spicker (eds.): *Euthanasia and the Newborn.* Conflicts Regarding Saving Lives. 1987
ISBN 90-277-2299-4; Pb 1-55608-039-5
25. S.F. Spicker, S.R. Ingman and I.R. Lawson (eds.): *Ethical Dimensions of Geriatric Care.* Value Conflicts for the 21th Century. 1987 ISBN 1-55608-027-1
26. L. Nordenfelt: *On the Nature of Health.* An Action-Theoretic Approach. 2nd, rev. ed. 1995 ISBN 0-7923-3369-1; Pb 0-7923-3470-1
27. S.F. Spicker, W.B. Bondeson and H. Tristram Engelhardt, Jr. (eds.): *The Contraceptive Ethos.* Reproductive Rights and Responsibilities. 1987
ISBN 1-55608-035-2
28. S.F. Spicker, I. Alon, A. de Vries and H. Tristram Engelhardt, Jr. (eds.): *The Use of Human Beings in Research.* With Special Reference to Clinical Trials. 1988
ISBN 1-55608-043-3
29. N.M.P. King, L.R. Churchill and A.W. Cross (eds.): *The Physician as Captain of the Ship.* A Critical Reappraisal. 1988 ISBN 1-55608-044-1
30. H.-M. Sass and R.U. Massey (eds.): *Health Care Systems.* Moral Conflicts in European and American Public Policy. 1988 ISBN 1-55608-045-X
31. R.M. Zaner (ed.): *Death: Beyond Whole-Brain Criteria.* 1988
ISBN 1-55608-053-0
32. B.A. Brody (ed.): *Moral Theory and Moral Judgments in Medical Ethics.* 1988
ISBN 1-55608-060-3
33. L.M. Kopelman and J.C. Moskop (eds.): *Children and Health Care.* Moral and Social Issues. 1989 ISBN 1-55608-078-6
34. E.D. Pellegrino, J.P. Langan and J. Collins Harvey (eds.): *Catholic Perspectives on Medical Morals.* Foundational Issues. 1989 ISBN 1-55608-083-2
35. B.A. Brody (ed.): *Suicide and Euthanasia.* Historical and Contemporary Themes. 1989 ISBN 0-7923-0106-4
36. H.A.M.J. ten Have, G.K. Kimsma and S.F. Spicker (eds.): *The Growth of Medical Knowledge.* 1990 ISBN 0-7923-0736-4
37. I. Löwy (ed.): *The Polish School of Philosophy of Medicine.* From Tytus Chałubiński (1820–1889) to Ludwik Fleck (1896–1961). 1990
ISBN 0-7923-0958-8
38. T.J. Bole III and W.B. Bondeson: *Rights to Health Care.* 1991
ISBN 0-7923-1137-X

Philosophy and Medicine

39. M.A.G. Cutter and E.E. Shelp (eds.): *Competency*. A Study of Informal Competency Determinations in Primary Care. 1991 ISBN 0-7923-1304-6
40. J.L. Peset and D. Gracia (eds.): *The Ethics of Diagnosis*. 1992 ISBN 0-7923-1544-8
41. K.W. Wildes, S.J., F. Abel, S.J. and J.C. Harvey (eds.): *Birth, Suffering, and Death*. Catholic Perspectives at the Edges of Life. 1992 [CSiB-1] ISBN 0-7923-1547-2; Pb 0-7923-2545-1
42. S.K. Toombs: *The Meaning of Illness*. A Phenomenological Account of the Different Perspectives of Physician and Patient. 1992 ISBN 0-7923-1570-7; Pb 0-7923-2443-9
43. D. Leder (ed.): *The Body in Medical Thought and Practice*. 1992 ISBN 0-7923-1657-6
44. C. Delkeskamp-Hayes and M.A.G. Cutter (eds.): *Science, Technology, and the Art of Medicine*. European-American Dialogues. 1993 ISBN 0-7923-1869-2
45. R. Baker, D. Porter and R. Porter (eds.): *The Codification of Medical Morality*. Historical and Philosophical Studies of the Formalization of Western Medical Morality in the 18th and 19th Centuries, Volume One: Medical Ethics and Etiquette in the 18th Century. 1993 ISBN 0-7923-1921-4
46. K. Bayertz (ed.): *The Concept of Moral Consensus*. The Case of Technological Interventions in Human Reproduction. 1994 ISBN 0-7923-2615-6
47. L. Nordenfelt (ed.): *Concepts and Measurement of Quality of Life in Health Care*. 1994 [ESiP-1] ISBN 0-7923-2824-8
48. R. Baker and M.A. Strosberg (eds.) with the assistance of J. Bynum: *Legislating Medical Ethics*. A Study of the New York State Do-Not-Resuscitate Law. 1995 ISBN 0-7923-2995-3
49. R. Baker (ed.): *The Codification of Medical Morality*. Historical and Philosophical Studies of the Formalization of Western Morality in the 18th and 19th Centuries, Volume Two: Anglo-American Medical Ethics and Medical Jurisprudence in the 19th Century. 1995 ISBN 0-7923-3528-7; Pb 0-7923-3529-5
50. R.A. Carson and C.R. Burns (eds.): *Philosophy of Medicine and Bioethics*. A Twenty-Year Retrospective and Critical Appraisal. 1997 ISBN 0-7923-3545-7
51. K.W. Wildes, S.J. (ed.): *Critical Choices and Critical Care*. Catholic Perspectives on Allocating Resources in Intensive Care Medicine. 1995 [CSiB-2] ISBN 0-7923-3382-9
52. K. Bayertz (ed.): *Sanctity of Life and Human Dignity*. 1996 ISBN 0-7923-3739-5
53. Kevin Wm. Wildes, S.J. (ed.): *Infertility: A Crossroad of Faith, Medicine, and Technology*. 1996 ISBN 0-7923-4061-2
54. Kazumasa Hoshino (ed.): *Japanese and Western Bioethics*. Studies in Moral Diversity. 1996 ISBN 0-7923-4112-0

Philosophy and Medicine

55. E. Agius and S. Busuttil (eds.): *Germ-Line Intervention and our Responsibilities to Future Generations.* 1998 ISBN 0-7923-4828-1
56. L.B. McCullough: *John Gregory and the Invention of Professional Medical Ethics and the Professional Medical Ethics and the Profession of Medicine.* 1998 ISBN 0-7923-4917-2
57. L.B. McCullough: *John Gregory's Writing on Medical Ethics and Philosophy of Medicine.* 1998 [CoME-1] ISBN 0-7923-5000-6
58. H.A.M.J. ten Have and H.-M. Sass (eds.): *Consensus Formation in Healthcare Ethics.* 1998 [ESiP-2] ISBN 0-7923-4944-X
59. H.A.M.J. ten Have and J.V.M. Welie (eds.): *Ownership of the Human Body.* Philosophical Considerations on the Use of the Human Body and its Parts in Healthcare. 1998 [ESiP-3] ISBN 0-7923-5150-9
60. M.J. Cherry (ed.): *Persons and Their Bodies.* Rights, Responsibilities, Relationships. 1999 ISBN 0-7923-5701-9
61. R. Fan (ed.): *Confucian Bioethics.* 1999 [ASiB-1] ISBN 0-7923-5723-X
62. L.M. Kopelman (ed.): *Building Bioethics.* Conversations with Clouser and Friends on Medical Ethics. 1999 ISBN 0-7923-5853-8
63. W.E. Stempsey: *Disease and Diagnosis.* 2000 PB ISBN 0-7923-6322-1
64. H.T. Engelhardt (ed.): *The Philosophy of Medicine.* Framing the Field. 2000 ISBN 0-7923-6223-3
65. S. Wear, J.J. Bono, G. Logue and A. McEvoy (eds.): *Ethical Issues in Health Care on the Frontiers of the Twenty-First Century.* 2000 ISBN 0-7923-6277-2
66. M. Potts, P.A. Byrne and R.G. Nilges (eds.): *Beyond Brain Death.* The Case Against Brain Based Criteria for Human Death. 2000 ISBN 0-7923-6578-X
67. L.M. Kopelman and K.A. De Ville (eds.): *Physician-Assisted Suicide.* What are the Issues? 2001 ISBN 0-7923-7142-9
68. S.K. Toombs (ed.): *Handbook of Phenomenology and Medicine.* 2001 ISBN 1-4020-0151-7; Pb 1-4020-0200-9
69. R. ter Meulen, W. Arts and R. Muffels (eds.): *Solidarity in Health and Social Care in Europe.* 2001 ISBN 1-4020-0164-9
70. A. Nordgren: *Responsible Genetics.* The Moral Responsibility of Geneticists for the Consequences of Human Genetics Research. 2001 ISBN 1-4020-0201-7
71. J. Tao Lai Po-wah (ed.): *Cross-Cultural Perspectives on the (Im)Possibility of Global Bioethics.* 2002 [ASiB-2] ISBN 1-4020-0498-2
72. P. Taboada, K. Fedoryka Cuddeback and P. Donohue-White (eds.): *Person, Society and Value.* Towards a Personalist Concept of Health. 2002 ISBN 1-4020-0503-2
73. J. Li: *Can Death Be a Harm to the Person Who Dies?* 2002 ISBN 1-4020-0505-9

Philosophy and Medicine

74. H.T. Engelhardt, Jr. and L.M. Rasmussen (eds.): *Bioethics and Moral Content: National Traditions of Health Care Morality*. Papers dedicated in tribute to Kazumasa Hoshino. 2002 ISBN 1-4020-6828-2
75. L.S. Parker and R.A. Ankeny (eds.): *Mutating Concepts, Evolving Disciplines: Genetics, Medicine, and Society*. 2002 ISBN 1-4020-1040-0
76. W.B. Bondeson and J.W. Jones (eds.): *The Ethics of Managed Care: Professional Integrity and Patient Rights*. 2002 ISBN 1-4020-1045-1
77. K.L. Vaux, S. Vaux and M. Sternberg (eds.): *Covenants of Life. Contemporary Medical Ethics in Light of the Thought of Paul Ramsey*. 2002 ISBN 1-4020-1053-2
78. G. Khushf (ed.): *Handbook of Bioethics: Taking Stock of the Field from a Philosophical Perspective.* 2003 ISBN 1-4020-1870-3; Pb 1-4020-1893-2
79. A. Smith Iltis (ed.): *Institutional Integrity in Health Care*. 2003 ISBN 1-4020-1782-0
80. R.Z. Qiu (ed.): *Bioethics: Asian Perspectives A Quest for Moral Diversity.* 2003 [ASiB-3] ISBN 1-4020-1795-2
81. M.A.G. Cutter: *Reframing Disease Contextually*. 2003 ISBN 1-4020-1796-0
82. J. Seifert: *The Philosophical Diseases of Medicine and Their Cure*. Philosophy and Ethics of Medicine, Vol. 1: Foundations. 2004 ISBN 1-4020-2870-9
83. W.E. Stempsey (ed.): *Elisha Bartlett's Philosophy of Medicine*. 2004 [CoME-2] ISBN 1-4020-3041-X
84. C. Tollefsen (ed.): *John Paul II's Contribution to Catholic Bioethics*. 2005 [CSiB-3] ISBN 1-4020-3129-7
85. C. Kaczor: *The Edge of Life*. Human Dignity and Contemporary Bioethics. 2005 [CSiB-4] ISBN 1-4020-3155-6

KLUWER ACADEMIC PUBLISHERS – DORDRECHT / BOSTON / LONDON